PROSTHETIC BODIES

Prosthetic Bodies

The Construction of the Fetus
and the Couple as Patients
in Reproductive Technologies

by

Irma van der Ploeg

Erasmus University, Rotterdam, The Netherlands

KLUWER ACADEMIC PUBLISHERS

DORDRECHT / BOSTON / LONDON

A C.I.P. Catalogue record for this book is available from the Library of Congress.

ISBN 978-90-481-5866-9

Published by Kluwer Academic Publishers,
P.O. Box 17, 3300 AA Dordrecht, The Netherlands.

Sold and distributed in North, Central and South America
by Kluwer Academic Publishers,
101 Philip Drive, Norwell, MA 02061, U.S.A.

In all other countries, sold and distributed
by Kluwer Academic Publishers,
P.O. Box 322, 3300 AH Dordrecht, The Netherlands.

Printed on acid-free paper

Printed in the Netherlands.

CONTENTS

INTRODUCTION

THE PARADOX OF REPRODUCTIVE TECHNOLOGIES

At the beginning of the twenty-first century, we live in a world that differs in many ways from the one of our ancestors. This book is about two contemporary phenomena, the existence of which these ancestors could not have anticipated, not even in their wildest dreams. The first concerns today's level of medical-technical control over human reproduction. The possibility of contraception, the creation of embryos in laboratories, or the temporary removal of fetuses from women's pregnant bellies are impressive technological feats profoundly affecting and altering fundamental human experiences. The second phenomenon defining our times, and equally constituting a radical change, is the level of emancipation and autonomy achieved by women. Never before have they been able to enjoy such relative freedom and opportunity to live their lives the way they want or to choose occupations, lifestyles, and partners. No longer are their lives and identities determined by duties toward husbands and children to the extent they once were, nor is the bearing and rearing of children the unquestionable first and foremost goal of their lives.

The rise of feminism and the progress of medical science are generally considered valuable achievements of our day and age. However, the relationship between the two is fraught with tensions. While there is, by and large, a broad consensus on their mutually reinforcing relationship in areas like contraception, it is far less clear how they relate to each other in domains like high-tech infertility treatment and prenatal medicine. When it comes to the new reproductive technologies, there is much more controversy, with doctors, feminists, patients, as well as concerned professionals from a wide variety of disciplines and religious or political bents pitched against each other. If there is one thing about these new medical reproductive technologies anyone can probably agree on, it is the fact that they have hardly been neglected or otherwise slipped through public attention. If anything, they have been discussed over and over, generating a steady flow of news items, legislative efforts, legal controversies, public moral upheaval, and academic attention from a wide variety of disciplines.

At the very moment this book goes to press, news media all over the world are covering the story of the "Italian madman" announcing his plans to try 'human cloning' – somatic cell nuclear transplant or SCNT – as a new infertility treatment. By now it is a near impossible task to give a complete overview of the instances of public controversy these technologies have generated over the last few decades.[1]

This book touches upon these debates in a sideways manner. Its main theme is the paradox, perhaps even the contradiction, that has resulted from the combined achievements of medical reproductive technologies and feminism. In significant ways, these technologies suggest a trend in which ever more medical and societal problems are becoming recast as reproductive problems, and ever more of those are finding solutions in interventions in women's bodies. Both in vitro fertilization and prenatal medicine are concerned with problems far surpassing those concerning women's own reproductive health and safety. In vitro fertilization has become important in managing the problem of *male* infertility, while prenatal medicine has yielded a form of surgery that is concerned with the prevention of congenital problems in *children*. Both types of technology comprise invasive, complication-ridden, sometimes risky procedures in women's bodies. They thus seem to constitute, on one level, invitations to women to take precisely the position regarding husbands and children that, on another level, so it was agreed, could or should no longer be demanded of them. In a certain way, these technologies invite women to take responsibility for men's and children's physical problems, sometimes at a considerable expense of their own well-being.

This way of framing the issue of reproductive technologies is not a very common one. There have been few protesting voices against broadening the indications of a treatment first introduced for women's fertility problems to those of men. On the contrary, the use of IVF for male infertility has been welcomed as a great medical breakthrough, and the waiting lists grow longer each day. And although many will still frown upon the idea of fetal surgery, the gradual extension of prenatal care and the concomitant growth in responsibilites, duties, restrictions and physical interventions for women has occurred without significant opposition. It is likely that, given this context, the inclusion of surgical options will eventually come to look like only one, logical step further.

If there has been little recognition of the paradoxical relation between women's recently achieved relative freedom *not* to subordinate their own well-being, interests and pursuits to those of husbands and children, let alone by jeopardizing such a highly valued good as their own health, and the directions taken by today's reproductive technologies, this has many reasons.

One such reason is precisely the extent to which women today are considered autonomous persons quite capable of determining what they want and need. An obvious counter argument to any suggestion that these technologies may be problematic from a feminist point of view is that women seem to want them. If IVF is rapidly becoming the treatment of choice for male infertility, this is because women do want children by their partners, and are prepared to go at great lengths to achieve this. A similar point can be made regarding prenatal intervention: it is because women put such great value on their children's health that they are willing to use all technologies available. If for some this will include undergoing surgery while pregnant, this will be an expression of their own free will to do anything possible to save their child. Questioning this choice could actually be taken as a sign of disrespect for their ability to make up their own minds.

Plausible as such reasoning may seem at first, this book will attempt to show that such an assessment rests on a superficial understanding of how and why we have ended up with these particular technologies and choices. To think that women *want* the painful and risky interventions in their bodies these technologies imply, because they want a child by their partner and one as healthy as possible as well, takes a great deal for granted. In particular, it assumes that these needs and wishes unavoidably and necessarily lead to, and therefore justify, current technologies. While it is true that most women would wellcome the long overdue lifting of the (cultural, including medical-scientific) taboo on male infertility, it is perhaps too convenient to take for granted, that they are therefore also happy to take its physical burden on themselves. It may be similarly true that women wish their children to be spared as much as possible from the painful strokes of fate of congenital disease and sickness, but it somehow seems unfair to infer too readily that this also means that they want or need their own already painful and risky task of childbearing to be increased indefinitely. Moreover, it is perhaps unjustified to assume that these translations of their needs were inevitable and necessary.

This book is an attempt to cast doubt on these assumptions, in order to broaden the scope of critical reflection about reproductive technologies as they are presently taking shape. Although some strong opinions and convictions may be read into this book, it does not offer general answers or blueprints. As a student of science and technology, trained in philosophy and women's studies, I use the modest instruments at my disposal. In writing this book I do not pretend to sit on the chairs of those whose works and words fill these pages. My position and equipment differ from those of doctors, scientists, ethicists, patients, policy makers, editorial boards, or distributors of research funds. I do hope, however, that those feeling addressed, from the various positions they occupy in relation to the technological developments discussed, will find some resources, arguments or suggestions helpful in evaluating, maybe even redirecting their agendas. In an effort to reframe some of the issues concerned, this book, in general, suggests a redirection of public concern raised by these technologies. It sets out to do so by bringing contemporary perspectives on the history of biomedicine and the nature of technology to bear on them. Through the analysis of the discursive practices of reproductive technology I will show how factors and mechanisms other than the natural, biological givens of bodies and reproduction are at work in shaping technologies that create the need for ever more interventions in female bodies for an ever growing set of reasons.

The first step in breaching the idea of biological inevitability is taken in chapter one, with a short account of the history of the gradual medical-scientific discovery and knowledge production about the female reproductive body. Next, the chapter introduces the theoretical and methodological approach to the study of science and technology informing the analyses in this study. It also provides a preliminary description of the two forms of reproductive technology to which these analyses are applied, IVF as a treatment for male infertility, and fetal surgery for the treatment of congenital anomaly.

Chapter two focuses the central argument on the emergence of two new types of patients, pivotal in the mediation of male fertility and congenital problems and female bodies: *the couple* and *the fetus*. It argues that the conception of couples and fetuses as singular treatable patients is closely connected to the development of reproductive technologies themselves. This constitutes a reflexive move in Katherine Hayles' definition of the term, according to which "that which has been used to generate a system is made, through a changed perspective, to become part of the system it generates."[2] Instead of describing the new patients as natural categories in need of therapeutical intervention, thus rendering intervention in female bodies biologicaly inevitable, this chapter shows their *production* through the intricate intertwinements and interactions of technologies and bodies. Focusing on the transformation of problem-definitions, and spatial and temporal shifts in localizations of the medical problems concerned, it describes how couples and fetuses emerge as *hybrid entities* from medical-technological interventions in female bodies.

Chapter three takes the analysis of couples and fetuses as hybrid patient categories one step further. It relates to chapter two as its twin sister, following a conceptual scheme devised by Bruno Latour.[3] According to this scheme, processes of hybridization are accompanied by processes of *purification*. While, within our technological culture, a continuous production of hybrids through the intermingling of nature and culture, bodies and machines, is taking place, this process is simultaneously concealed by discursive reconstruction of pure and distinct ontological categories. The very culture that so habitually and productively mixes the natural and the technological, thus maintains an ontology in which nature and technology, bodies and machines are fundamentally distinct. A similar scheme is used to analyze the relation between couples and fetuses on the one hand, and reproductive technologies and the individuality of bodies on the other. While chapter two deals with the production of fetuses and couples as hybrids, chapter three centers around the purification processes involved. The central argument of the third chapter, then, is that although couples and fetuses are treated as singular patients, these unconventional practices, that mix up the distinction between the technological and the natural as well as the distinction between one individual and another, are accompanied by specific discursive mechanisms that render the "impure" categories familiar and acceptable. A crucial step in rendering the treatment of male problems and children's problems through women's bodies acceptable, or even natural and biologically inevitable, lies in a discursive displacement of 'women' by 'couples' and 'fetuses', and the subsequent reconstruction of the new technologies as forms of treating 'men' and 'children'. The chapter describes how the scientific discourse concerned presents the new technologies as hardly involving any intervention in female bodies, as opposed to the couple's or the fetal body, while, on the other hand, the individuals treated and helped through the technology are referred to as men and children.

Chapter four, then, poses the question what these configurations imply for the female body: if the technology applied to it restores men's and children's bodily functions, what remains of the female body as an *individual* body? What remains of the idea that bodies have *boundaries* marking the distinction between one body and another? These questions about the body ontologies produced within technology's discursive practices are extended to the domain of moral and legal notions of the individuality of the body in relation to concepts relevant to body politics and reproductive politics, such as bodily integrity and bodily self-determination.

Chapter five, finally, answers the central question of this book by drawing together the conclusions from the previous chapters. The question how to assess current developments in reproductive technologies is answered by comparing technology's constitution of women's relations to men and children through its rendering of forms of female embodiment with feminist goals and emancipatory body politics. It then moves the problem to a different plane of discussion by confronting these conclusions with current developments in *feminist theory*. In its emphasis on dissolution and contingency of boundaries and concepts like the self, the individual and the body, postmodern feminist theory seems to have discarded the very notions that provided feminism its cultural and political leverage and possibilities to critically assess technological developments. In this epilogue on the theory of science, technology and the body I argue for a postmodern feminism that acknowledges both the contingency and indispensibility of modern ideals of women's individuality and bodily integrity.

CHAPTER 1

PRELIMINARY MOVEMENTS

The Body of Theories, Practices and Texts

1. INTRODUCTION

The first step to be taken in loosening the apparently self-evident links between women's wants and needs and the particular answers offered by contemporary technologies concerns addressing the underlying assumption of their biological inevitability. There is a persistent belief in the status of biomedical knowledge, the supposed basis for current technological practices, accountable in considerable part for this sense of inevitability. If there is no sense of any paradox arising from women's emancipatory efforts and current technological developments, this has to do with a widely held presupposition that the very notion of politics is not applicable when it comes to bio-medicine. Biology, describing the *nature* of bodies, cannot be political; therefore, as long as biological knowledge is considered to be true (and as long as the technologies designed on that basis, as the standard view on these matters assumes, work), it cannot be contested; biological knowledge merely forms a growing set of ahistorical, natural facts, that we can only discover, accept, and use to our advantage. In the context of reproductive technology, this broad and general belief translates into the following one: if women's bodies are the object of intervention in practically all reproductive technologies, and for medical problems that by long have surpassed those concerning their own reproductive health, this has nothing to do with any kind of politics. It merely reflects and follows from the biology of reproduction. There is no way around the fact that it is women who have children, and from this everything else follows. Women can freely choose to take or leave the technologies on offer, but the configuration determining the choices open to them has been shaped by the biology of reproduction itself.

In order to begin critical reflection upon our culture's apparent need for the technologies it has developed, it is this idea of biological inevitability that needs nuancing. If it is the nature of the female body that dictates how and where medical problems surrounding reproduction are to be located and addressed, this is a nature that resulted from a very particular history. The next section sketches patterns in the history of medicine's and science's dealings with the female body, and the resulting knowledges and practices, and indicates in a provisional manner how these patterns continue to play a role in the present.

Next, and this constitutes the main challenge of this book, it is crucial to demystify the notion that the new reproductive technologies provide the definitive and inescapable singular answer to women's problems and desires. This requires a perspective on technology that differs from the one that is usually implicit in public reflections and evaluations of reproductive technologies. Common to most conceptualizations of the issues arising from these technologies is their framing in terms of *effects* and *consequences*, or even of consequences and effects of certain *applications* of technology. This pattern is probably largely accountable for the fact that of the many feminist concerns about these technologies, only those restricted to debating the risks, the efficiency, and, occasionally, the "psychological impact" of these technologies, succeeded in gaining a wider hearing. The same pattern underlies the emphasis in many public and political debates on these technologies on ethics. A frequently used way to raise public concern about technology is a statement about there being "moral issues involved", as a sort of appendix to the mere technical, medical aspects.[4] How exactly the implied boundary between fact and value is drawn may vary according to one's preferences in what should be up for debate, but the distinction is usually there.

But wherever the line between the technology itself and its consequences or effects, between the methods and their applications may be drawn, using such distinctions presupposes that there is a sphere where technology, science or medicine exists in a pure, neutral form. So, paradoxically, this way of defining public problems concerning medical science and technologies actually constructs them in a way that simultaneously puts them beyond the grasp of moral or political scrutiny to considerable extent, since the definition of the 'moral problems' is postponed until after the establishment of 'the facts'. Hence the feeling of many that such reflections are always more or less running behind the facts and not really capable of influencing technological developments. Medical science and its inventions, so it is experienced, will always be one step ahead, with 'society' always reacting to the latest development, after the fact. This way of conceptualizing technology in effect creates a space where it can develop relatively undisturbed.

This book proceeds from a perspective on technology that is conceptually rooted in a type of technology studies that tries to focus on the "inner workings" of medical science and technology, as opposed to their effects, consequences and applications. It seeks to locate the moral and the political within what counts as 'technology itself', within what counts as 'scientific fact'; more precisely, it does not accept the distinction between technology and science on the one hand, and the moral, the political or 'external effects' on the other, as a priori given. Try, for instance, to explain or describe what the technology of IVF "in itself" is. There is no way to do this without describing what is *done*, and thus at least implicitly, such a circumscription contains a practice, some application, purpose or consequence; a norm for what counts as successful or standard application; it necessarily involves describing actions, patients and their body parts, and what happens to them. One cannot identify these technologies apart from what are usually thought to be 'external' aspects. This is especially true for complex technologies like IVF that involve a series of actions, techniques, machines and experts, so that there is no one particular machine or piece of hardware that might be (erroneously) identified with the 'technology itself'. Medical technology always implies a particular way of *doing* things, a *practice*.[5] Actions, attitudes, words, texts, values, norms, and social relations are considered as integral to the technology as the instruments, chemical substances, and laboratory procedures. The third section of this chapter elaborates this view on technology and its relation to science, and introduces some of the key concepts used in this study.

This chapter's fourth section introduces the two technological practices that form the subject of this study. Some "figures and facts" of in vitro fertilization, specifically as a treatment for male infertility, and fetal surgery are given, in order to delineate the contours of both practices. The final section develops the central questions of this study and discusses the theoretical views informing the analyses presented in this book. In particular, it argues how analyzing medical-scientific texts can contribute to our understanding of current developments in reproductive technologies as well as the way female bodies are configured in these practices.

2. A COINCIDENCE OF MEDICAL SCIENCE, HISTORY AND BODIES

A strong and widespread conviction persists, that from 'the rise of modern science and medicine', there has been a steady accumulation of empirically grounded, valid knowledge, that could not have been otherwise, since this knowledge steadily uncovered how bodies and reproduction actually work. At the same time, however, anyone with only a marginal interest in the subject will be aware of medical practices and theories concerning women (but not only them) in the past (but not only in the past) that from today's perspective are ranging from the laughable to the deeply shocking and apalling. But however such past practices and theories may be denounced today, they never seem to be considered of actual consequence for our knowledge and practices today, whether this 'past' concerns the early nineteenth century or a mere decade ago. Thus a strange combination of beliefs predominates views on contemporary reproductive medicine: on the one hand modern medicine and biomedical science are seen as long standing, cumulative traditions, while on the other hand its results and products, its theories, practices and technologies, are held to be untainted by anything so mundane and contingent as history and tradition.

It is not my intention to give here a comprehensive and detailed account of the history of modern science and medicine concerning sex, gender, and reproduction. I do want to draw attention, however, to some aspects of this history in as far as this provides some elementary and necessary background for the issues addressed in this book. As will become clear below, recent historiographical studies suggest that the more problematic historical aspects still thoroughly shape the reproductive biology that today is thought to dictate the configurations in which the female body appears as the object of interventions for an ever growing set of medical problems. To see this, one has to be prepared, if only for a moment, to postpone taking recourse to the ever available and all too often invoked male and female bodily differences and givens as explanations for the current nearly exclusive involvement of women in reproductive technologies.

This section is primarily an argument against biological determinism. In an attempt to move away from biological determinism, two main approaches have become prevalent. One type of criticism of science, developed mainly within women's studies of science and the philosophy of science emphasizes the cultural and social origins of the prejudices and biases that color the content of scientific knowledge. This critique has yielded a lot of convincing evidence against science's claims of neutrality and objectivity, on the levels of its choice and formulation of research questions[6], its methods[7], up to its epistemology and specific ideals of neutrality and objectivity[8]. Valuable as this type of criticism has been, it runs into some serious problems. In its conceptualization of 'biases' and 'prejudices' as the main causal factors in scientific knowledge production, it implicitly assumes that the ideals of neutrality and objectivity themselves still hold. Moreover, in attributing what it sees as 'bad science' to massive cultural and social factors, like gender structures and enduring male dominance, it replaces biological determinism merely by another type of monolithic determinism: a social or cultural determinism that reads too much intentionality and monocausality into the history of science. When "male thinking" and psychology are designated the main causal factors in the generation of biased knowledge, it even is in danger of letting biological determinism seep back in through the back door.

Against such readings of science argues a second approach that emphasizes the contingency of scientific knowledge production. Usually referred to as 'science and technology studies (STS)', this approach combines ethnography, sociology, historiography and philosophy in empirical studies stressing the heterogeneity of factors at work in scientific practices.[9] Rather than seeing scientific development as driven by (bad) ideas and broad, almost ahistorical cultural patterns, this tradition emphasizes institutional, social, and material factors that shape the specific, historical configurations of scientific work. Yet the problem with this approach - besides its rhetoric of 'empirical correctness' on account of the empirical detail it strives for - is that it tends to generate its own blind spots. Despite its claim to empirical comprehensiveness, it cannot avoid being selective as well. A bias toward classic sociological factors, such as institutions and interactions between groups, may have given way to a new trend of stressing 'material' factors and the role of artefacts and objects, but the selectivity necessarily remains. Thus, for example, in its focus on 'contingency', its identification of relevant factors in the construction of certain scientific facts or technologies may diffract in all kinds of directions in a particular episode. this may result in a failure to account for more enduring patterns over time.[10]   Both approaches are nevertheless important for their critique of scientific rationalism and biological determinism. The following reconstruction of developments in medical science, as pertaining to the historical background of contemporary knowledges and practices regarding reproductive bodies, makes use of indispensable insights from both traditions. As such it tries to avoid the pitfalls of overemphasizing intentionality and broad cultural determinisms on the one hand, and too much contingency and lack of awareness of more enduring patterns on the other.

Today there are myriads of ways, opportunities and reasons to intervene in women's reproductive bodies. Libraries are stacked with gynaecological atlases and textbooks, and we have an endlessly proliferated nomenclatura for potential female pathologies and conditions in this area of medicine. This situation results from a long tradition of medical and scientific practices aimed at the female reproductive body. In comparison to our knowledge of and attention for the male reproductive body, one could argue that women as reproductive bodies suffer from overexposure.

The development and production of knowledge about any phenomenon are not determined by intrinsic properties and characteristics of that phenomenon (since these are the very product, or substance of the resulting knowledge in question), but are rather a function of the extent and ways in which that phenomenon is problematized, i.e. defined and researched as a scientific or medical problem. One cannot expect substantial insight in male reproductive pathology, for instance, if there is not first a willingness and effort to problematize, analyse and experiment with male reproductive functions. What we are faced with today is the result of centuries of asymmetrical distributions in medical scientific interest and experimenting zeal regarding male and female reproductive functions.

The scientific practice of anatomy in the eighteenth century, usually identified with the crucial transition to a modern, empirically based medical science, came into being at a time that saw the discarding of an age-old, metaphysically grounded cosmology. In this cosmology, in which everyone had their natural place in accordance with a natural hierarchical order of being, no distinction was made between the natural and the social. This very distinction was in fact produced in the process of overthrowing this old cosmology (including its claim to "natural", god-given authority for some over everyone else) and resulted in the creation of a separate domain called "nature" over which church nor king would have any authority. This separation enabled the idea of an objective scientific study of nature, and the endowment of its results with the status of universal truth outside the scope of human or divine authority, politics, and prejudice.

A rapidly growing body of historical research shows that in this historical period, a mutually reinforcing relationship existed between the agenda anatomy set for itself and contemporary political struggles. Schiebinger (1993), for instance, describes how, precisely when Enlightenment's formulations of 'equality among all men' generated claims for political equality from women and people of color from the colonies, anatomists set out to investigate and locate sexual and racial difference in the body in historically unprecedented ways.[11]

Until then, male and female bodies were seen as essentially the same, with female reproductive organs as inversed versions of the male's. Sexual difference, as extensively described in Lacqueur (1990), was conceived as differential positioning of women and men on one scale of metaphysical being, with the male as the perfect form. Women were less perfect versions of males, caused by their lack of 'heat', a characterisic that, among other things, prevented their reproductive organs to extrude, as do the male ones. Thus, women were seen as producing seed, as men do, in their internal 'testes'; their vagina, cervix, and uterus were pictured as inversed versions of the penis and scrotum.[12] This account, especially in its formulation by Galen (2nd century A.D.), was passed on to medieval Western Europe through the writings of the Arab Ibn Sina (Avicenna), and remained authoritative and influential all through the Renaissance. It is, for instance, still discernible in the first anatomical drawings of dissected corpses by Da Vinci and Vesalius.

The newly articulated universal principle of equality among all men, however, provided no ground to exclude groups that were never really meant by the brotherhood of men to be included. In revolutionary France of the late eighteenth century, claims to the 'natural rights of man' by women and by people of color from the colonies were countered, on occasion quite explicitly, by the suggestion that 'natural', i.e., bodily, differences provided the justification of exclusion of some social groups from the "universal" political rights.[13] Practically at the same time, science began to 'discover' such differences.

Within this project, the anatomical search for sexual difference concentrated on female reproductive organs, and also on some other anatomical features, culturally associated with gender-roles, such as skeletons, in particular pelvises and skulls.[14] Generally speaking, women were studied to find out what distinguishes them from men. It was now that the ancient model of women-as-inversed-men gradually became replaced by the naturalists' complementarity-model of the sexes. This fundamentally new account saw the sexes, although each perfect in itself, as physically essentially and incompatibly different, instead of (im-)perfect versions of each other.[15]

It is highly significant that this shift did not coincide with the turn toward the empirical study of the human body by actual dissection and graphical description. The fact that the model of antiquity, with women's bodies as inversed, less perfect versions of men's bodies, remained in place for a long time, even after the first anatomists set out to dissect and study corpses, challenges the notion of objectivity as unmediated vision and pure observation as the distinguishing methodological novelty of modern, empirical anatomy. Corpses had been dissected occasionally for centuries, and the graphic rendering of female and male organs by Leonardo da Vinci and Vesalius show how they actually *saw* what they dissected as inversed versions of each other. This implies that it took something more than just opening the body and merely "looking how things really were", as the standard story about the "discovery" of the modern anatomical body goes. In addition to the relationship between the broader political changes mentioned and the definition of research questions and conceptualizations of difference, there are very practical and material factors determining what the eye can see at any particular historical moment. The 'anatomical gaze'[16] was shaped to a large extent by the development of specific tools and techniques as well. One cannot 'see' the anatomical body as it developed from the seventeenth century onwards, in messy and bloody bowels lying on a dissection table, with a naked, untrained and unaided eye; it took the later development of techniques of preservation for the corpses, and specifically drawings and etchings of the body, in order to guide and train the hand how to cut, and the eye what to see[17]. Evidently, the anatomical body produced by Enlightenment's pure empirical observation was very much a body crafted by a broad variety of historical factors.

The anatomical racial and gender characteristics produced were without exception conceptualized as difference from the white, male body - the norm against which all difference became essential alterity. With the nineteenth century rise of evolutionary models of human origins, alterity as 'race' became conceptualized as relative proximity to animals, most notably apes. White women, on the basis of their skulls being relatively larger than men's, were described as being less developed and closer to children. In searching for the boundary between humans and animals, male apes were studied to highlight potential distinguishing characteristics such as reason, speech, bipedalism, with no reference to sex. By contrast, when females were studied, only sexual traits were considered. According to Schiebinger, it was a general characteristic of European scientific studies since Aristotle that females, human or animal, were only studied insofar as they deviated from the male[18]. In nineteenth century debates on the distinction between humans and animals, female sexual organs were studied to highlight the animal side of human life; where human uniqueness was considered in females it came to focus on key sexual traits such as menstruation, the clitoris, the breasts, and the hymen. These naturalized, 'scientifically discovered facts' about "the great chain of being", with women and people of color positioned as evolutionary lower, became part of political debates about, for instance, political representation and access to scientific education.

It is hard to say whether this singular scientific focus on women as reproductive and sexual bodies followed from or resulted in the notion of femininity that dominated most of nineteenth century medical science. Woman's identity, or the essence of womanhood, so to speak, was thought to be determined by her reproductive functions. While the uterus preceded the ovaries in being taken as a "pars pro toto", the sentence "Propter solum ovarium mulier est id quod est (it is only because of the ovary that woman is what she is)[19] typically expresses the predominant nineteenth century view on the relationship between reproductive body parts and the "essence of woman". Based on this notion, her entire physical, mental and moral well-being was thought to be connected to her sexual and reproductive functions. There has not been a comparable emphasis on men's reproductive functions with respect to their general well being and functioning, because men, the rational sex with the exclusive talent to control nature, were not supposed to be similarly governed by their sexual physiology as were women.

It is specifically the limitlessness of the range of afflictions and pathologies that nineteenth century medical science associated with female sexual physiology that gave medicine the negative image it acquired in early works of contemporary feminist scholars. Specifically the new specialty of gynaecology, that claimed to study "the whole woman, fusing physical, psychological and moral aspects of femininity",[20] while limiting itself to the reproductive organs, has been the subject of quite critical forms of historiography. The way in which mental and behavioral deviancy, and even female attractiveness and moral worth, were connected to reproductive physiology (most notably, the ovarian function) and turned into reasons for medical intervention and surveillance, has fed theories on medicine's infamous role in proscribing and sanctioning narrowly defined gender roles and double moral standards. The idea that femininity practically equalled frailty and proneness to pathology found its way, for instance, into convictions about the debilitating effects of menstruation on mental capacities and medical warnings against higher education for women[21]. Many were the health hazards women were seen liable to when stepping out of the narrow boundaries of their assigned gender roles.[22] Restrictive ideals of femininity had been so evidently shaping medical views that it is hardly surprising that many of the first studies uncovering this history saw medicine and gynaecology as male controlled instruments of oppression.[23]

More recent work is nuancing these first harsh judgements, by focusing, for instance, on the controversies among medical professionals in the periods studied, thus countering any monolithic picture of past practices. Moscucci has shown, for example, that the practice of ovarectomy - one of the standard examples of medical "abuse" of women for what are now seen to be very unsound medical reasons - generated from the beginning fierce opposition within a strongly divided medical community, causing heated polemics and debates up into the highest strata of medical professional organizations.[24] The reasons, motives, and results for this high risk experimental procedure were contested by nineteenth century medical professionals in terms that make the fiercest feminist condemnations in the twentieth century of this practice seem almost friendly in comparison.

But even these corrective historical accounts still highlight overt differences in medical treatment of men and women that cannot be accounted for in scientific, medical or biological terms, and that should not be ignored when trying to account for technological configurations today. With respect to the contrast between the willingness, or perceived necessity to intervene in women's reproductive bodies, with drastic, highly experimental procedures, and the absence of such practices regarding male reproductive bodies, combined with the scale on which such practices took place, the practice of ovarectomy remains a highly relevant case in point. Controversy notwithstanding, by 1870, it had become an accepted practice, long before there had been any significant drop in its mortality rates, which was often well over fifty percent[25]. It was performed for a wide variety of reasons, most of which were non-life-threatening and involving healthy ovaries, such as in cases of behavioral pathologies, dysmenorrhoea or even prevention of pregnancy.[26] There has not been a practice equivalent to what actually amounted to female castration for males: removal of healthy testes never became an accepted intervention, a few cases of criminal insanity and cancer of the prostate excepted[27]. Male pathologies of the reproductive system were not at any time seen as calling for such draconian experimental measures, let alone the scale at which they were applied to women. Significantly, gynaecological experimental surgery, at the time, was considered learning experience for surgeons, providing the medical knowledge and skill for carrying out abdominal surgery generally. At the first conference of the British Gynaecological Society in 1891, gynaecological surgeon James Murphy stated for instance: "Ovariotomy has opened up the whole field of abdominal surgery, so that many men who started as gynaecologists are now our most brilliant surgeons, successfully attacking the uterus, the spleen, the liver, and all the other organs contained in the abdomen."[28] Finally, and perhaps most importantly, the operation was performed countless times before any scientific consensus existed on the question what effect removing ovaries would actually have[29]. This knowledge was, in the end, what *resulted* from thousands of experimental surgeries and many deaths.

When moving into the twentieth century, with its endocrinological revolution and concomitant changes in medical views on reproductive physiology as well as new therapeutical emphases, these nineteenth century patterns and practices can be seen to have prestructured newly emerging practices in ways that reproduced similar asymmetrical treatments of the sexes. Oudshoorn (1994) has described how the study, making, and marketing of sex hormones has come to concentrate so much on the female sex.[30] Though there was initially as much interest in the study of the role of sex hormones in the male body and in the process of sex differentiation itself, it did not take long before the research on sex hormones was predominantly linked to the female body and pathologies of the female reproductive system, reducing the other research interests to marginality. Mainly relying on a sociological perspective, Oudshoorn has convincingly shown that this was largely the effect of the links that were set up between gynaecological clinics, laboratories, and pharmaceutical companies. The gynaecological clinics provided the other two parties with access to large patient populations (being test populations, providers of 'raw materials', as well as the targeted consumers), research materials in the form of ovaries and urine, enormous quantities of which were needed to isolate the wanted substances, as well as research questions foregrounding issues connected to female reproductive functioning. Such a link could not be established with male oriented clinics, for the simple reason that there was (and is[31]) no medical discipline or comparable practice for the male reproductive system. With regard to the necessary research materials, there were no quantities of testes available comparable to the amounts of excised ovaries, because, as we have seen, the male gonads were not seen as inherently pathogenic for "the whole man" as they were for women. This latter aspect had further consequences for the subsequent marketability of sex hormones as therapies, something which pushed the involvement of pharmaceutical companies to a substantial extent. In line with prevailing ideas about women being naturally prone to all sorts of pathologies and disabilites due to their reproductive system (for men there was hardly such a connection thought worthy of investigation), the "new drugs looking for diseases"[32] found a wealth of potential indications for women, while there was no such fertile ground to receive male sex hormone therapy.

Against a background of sky-high expectations concerning their profitability because of their expected general applicability to all 'women's diseases', up to "stimulating femininity and beauty"[33] itself, sex hormones were marketed from the early 1920s on a fast growing scale, even before clinical test results were available. Notwithstanding this lack of what even at the time was considered a sound 'scientific basis', by the early 1930s, female sex hormone therapy was prescribed for all menstrual disorders, all anomalies of ovarian function, sexual function, menopause, infertility, problems of the genital organs, psychiatric disorders such as schizophrenia and melancholia, psychoses and depressions, dermatological diseases, diseases of the joints, epilepsy, hair loss, eye disorders, diabetes, hemophilia and chilblained feet.[34] At the same time, a clear 'clinical picture' for female sex hormones was still absent. Therefore, large scale clinical trials were set up to investigate their therapeutic activity in women[35]. The first reports about carcinogenic effects of female sex hormone therapy also appeared in 1932, but this failed to have a noticeable effect on promotion or reception.[36] Although such reversed relations between large-scale introduction and availability of what can count as an "empirical basis" are not unique to this case, in the case of male sex hormone therapy it worked exactly the other way around: with the absence of any such high expectations about universal beneficence in men (corrolary to the absence of beliefs in universal deficits in men), the lack of a clear picture of activity of sex hormones in males, was frequently put forward by pharmaceutical companies as well as medical professionals as the reason for extreme cautiousness in marketing and suggesting applications (eventually only a few urological disorders).[37]

It appears, then, that the pattern of a long standing lack of medical-scientific interest in the male reproductive functions, in the problematization of its functions and pathologies, tends to reproduce itself. The consequent absence of a male counterpart to gynaecology and the practice of ovarectomy, pre-structured the new and apparently unconnected medical-scientific field of endocrinology, and eventually reinforced the assymmetrical pattern. This is also true with respect to the relationship between wide-scale introduction of an intervention in practice and availability/production of empirical knowledge about its effects. Before anything near a significant scientific agreement on the functioning of hormones was reached, they were prescribed to thousands of women, for countless reasons, yielding in the end - and with the stories of the damage done along the way largely gone to dust[38] - an abundance indeed of empirical knowledge about the role of hormones in the female body. By contrast, application and experimentation on men was negligible, and hence, a negligible knowledge production or therapeutical arsenal resulted.

Throughout the second half of the twentieth century, these patterns were repeated in various instances similar divergences and asymmetries in the research and treatment of the sexes, for example, in the development of contraceptives and the hausse in hysterectomies in the 1970s and 1980s - the twentieth century version of the ovarectomy-story of the nineteenth century, this time on an even much larger scale.[39]

Enduring as well appears the negative language and imagery medical science uses to describe the female reproductive body. Connected to the epidemic in hysterectomies, for instance, was the then prevailing image of the (non-pregnant) uterus as a useless organ, a liability best to be gotten rid of.[40] In accounts of physiology and anatomy in medical literature, hospital manuals, and school textbooks, negative connotations have been shown to abound when the subject turns to female reproductive physiology.[41] Normal physiological processes like menstruation and menopause are being described in terms connoting failure, breakdown, deficit and functionlessness; labor and birth in terms of mechanistic production, with the inefficient uterus-machine prone to countless dysfunctions; and pregnancy and birth as a foetus' journey through dangerous territory and hostile environment. With such a medical-scientific rendering of female physiology, women's bodies are still all too often appearing as more 'naturally' in need of technological intervention. Today, many women are becoming convinced that they cannot age safely without decennia of hormonal replacement therapy,[42] and recently it has been suggested that they cannot mature safely without similar treatment from puberty on.[43] Next to ovaries and uteruses, breasts have, with the recent discovery of the "breast-cancer gene", become a liability that may be better amputated preventively, just as ovulation, according to some, may be a mortal danger from which women are to be saved by putting twelve-year-olds on the pill.[44]

It is safe to conclude that a historical pattern persists, leaving us with an ambiguous inheritance. The focus on women's reproductive and sexual parts and functions has caused a far-reaching medicalization of women's lives, an abundance of detailed knowledge and a wide array of the most sophisticated therapies and technologies (a mixed blessing in itself). Yet the other side of the coin is that health problems *not* related to sex or reproduction have become under-researched in women. In these areas, research is often conducted on male bodies only. Females are regularly excluded from clinical trials and animal models because of the "confounding effects" of their bodies' peculiarities on the results.[45] The results are nevertheless generalized to women, gender differences in etiology, symptoms, and reaction to medication notwithstanding. Research into cardiac disease, AIDS, various types of cancer and occupational exposure to chemical substances are major examples of this pattern.[46]

By contrast, the *under*exposure of men as reproducing, sexual beings in medical and scientific research has left medicine comparatively empty-handed in these areas to the detriment of men's reproductive health as well a that of women. The mirror image of the notion that female sexual organs cannot but cause trouble and disease is the equally damaging notion that nothing really and hardly anything can be wrong with male reproductive functions. Thus, for example, it has taken far too long to acknowledge that occupational exposure to certain chemical substances is not only dangerous to male individuals themselves and to women in as far as they are potential reproducers, but actually can cause severe gonadal damage and infertility in men, and, through them, miscarriages in their female partners, and congenital disorders in their children.[47] Underdiagnosis of male sex and reproduction-related health problems is problematic from the point of view of men's health; the consequent misdirected and repetitive overtreatment of their female partners has been damaging to women's health. Such neglect is rather serious, but becomes even more so when what is presented as compensation and amendment of longstanding, lopsided developments and practices (as IVF and its extensions, for instance, are often seen as finally addressing the neglected problem of male infertility) is once more premised on interventions and experiments on women's bodies rather than men's.

It needs perhaps to be stressed once more that all of the above has not been recounted here to impute medicine or science, doctors or scientists, of bad intentions and victimization of women. The *causes* for the way knowledges and practices have developed were pluriform and have only been tangentially mentioned here. If ever one must emphasize the heterogeneity of factors at work in steering developments into particular patterns, it is here, where obviously cultural, economical, social, institutional, psychological and historical elements come together, in various, regularly changing configurations.

On the other hand, recurring patterns *are* discernible. This must give pause to theories that tend to overemphasize contingency in the development of science and medicine. However contingently, and perhaps questionably a certain fact, theory, or technique may have come about, once established and elevated to hard scientific fact or accepted medical practice, that is, once its origins are forgotten, it becomes the invisible foundation of future developments - one that cannot be undone as easily as one might hope. Descriptions of the genealogy or the critical historiography of specific scientific beliefs or medical practices by themselves cannot make a change nor constitute refutations. Apparently, there is much more continuity in scientific developments than theories of radical contingency sometimes allow for. A focus on gender provides one way to uncover persistent patterns in medical-scientific research and clinical practice, and, considering the unhealthy nature of the patterns brought to light already, a most important one as well.

No matter how these patterns may have come about, and to whatever contingent factors they may be attributed, it is above all the form of these patterns that provide ample reason to question basic notions of scientific objectivity, rationality, and neutrality. This seems to have particular relevance when it comes to issues of gender and sexual and reproductive matters. The suggestion that the shape of our current knowledges and practices cannot be simply attributed to nature or biology - the way things 'inevitably' are - should be warranted more attention. Moreover, as long as there is little recognition that such patterns exist, and that they are not merely curious things of the past, we should perhaps expect these patterns to reproduce themselves. They still shape our current practices and technologies by virtue of their character as unnoticed black boxes, the best intentions of contemporary actors notwithstanding, and however 'revolutionary' and 'new' contemporary developments may seem.

3. TECHNOLOGY AND THE TRANSFORMATION OF PROBLEM-
DEFINITIONS

This chapter began with a discussion of the paradox implied in contemporary reproductive technologies. I pointed out how these sophisticated and contemporary technologies are inviting women to take up positions and responsibilities regarding husbands and children highly reminiscent of the ones society - as a result of feminism - had finally ceased to demand of them. But instead of even acknowledging the existence of this paradox, these technologies are often exempted from criticism and merely viewed as something women themselves ask for. The fact that women resort to these technologies is even seen as an expression of their emancipated, free and autonomous choice, whereas any suggestion of the complexity of the relationship between new reproductive technologies and feminism tends to be easily dismissed.

The relatively young field of science and technology studies provides several new insights and concepts that enable us to question too hasty equations between women's needs and desires and the current technological configurations. It opens up venues toward a better understanding of the ways in which current technological practices tie together histories of knowledge production, revolutionary techniques and instruments, medical problem-definitions, and ways of configuring bodies. It is specifically the concept of 'translation' that promises to be helpful, for instance, if one wants to understand how children's congenital diseases and men's fertility problems have come to be defined within the context of reproductive technologies as problems indicating the need for invasive medical treatment of women.

In its efforts to account for science and technology's effectivity and power, one strand of constructivism has elaborated this concept of translation as a key mechanism in the production of scientific truth and technological effectivity. According to Callon and Law[48], two of its major theorists, one of the chief operations of experimental science is to design a series of reformulations of a problem (that may also have a theoretical, societal or already medical formulation), so that it becomes amenable to analysis and experimentation on laboratory scale. There is usually a large distance between the concrete, detailed, technical events and actions carried out in the laboratory and the problems and questions such experiments are supposed to be (part of) the answer to. This distance is bridged by long chains of theoretical, conceptual as well as material reformulations. It is the extremely effective way in which science accomplishes this reformulation of problems (reshaping practically any problem in its own terms), that accounts in part for science' and technology's succes. Once a problem has found a scientific formulation, that is, once it has been translated into a form that renders it researchable in a laboratory, it acquires features and aspects that can be experimentally addressed, managed and manipulated.

One problem with this process, however, is that such reformulations are never clean, literal translations but always entail - no matter how subtle - a trans*form*ation of the problem. Feeling often tired and weak, for example, is not exactly identical to the problem of having a low hemoglobine level,[49] although treating it as such can have very beneficial effects. The first formulation figures in a different setting and is connected to a set of different meanings, actors, experiences and potential solutions. But on some occasions, treating the second problem may solve the first one. It is this changed aspect of the problem that accounts both for science's (or medicine's) success in solving problems, for it is through its unique capacity of transforming problems into "do-able",[50] manageable, and manipulable ones, that science often succeeds in identifying and designing effective interventions.

Simultaneously, however, this capacity to transform problems through reformulation, is also what accounts, in a nutshell, for science's Janus-faced character, as it is widely experienced today. Next to providing (half of) the basis of its success and power in society, this capacity of transforming problemdefinitions is precisely what feeds the often expressed discontent with the dominance of technical and scientific (or medical) approaches to so many societal and individual problems. For in order to be able to effect changes outside the laboratory, science has to convince others that it really has addressed the same problem. To ensure that its solutions will work in the "real world" as well, it is the "real world" that more or less has to conform to science's definition of what the world is like. For its laboratory-found solutions to hold in the real world as well, the real world has to become a bit laboratory-like[51]. Laboratory set-ups constitute configurations in which particular actors, objects, procedures have to stand in specific relationships to each other, so that, say, particular variables can be identified, of which some can be held stable, others be varied, certain compounds be added, procedures be executed, other aspects be measured, or visualized and so on. In order for an intervention designed in such a setting to work elsewhere as well, this entire configuration has to be reproduced and represented there as well. The intervention will not work if it is not made sure that the problem will let itself be transformed into the laboratory definition of the problem, and this consists, by and large, of actively bringing this transformation about. In a way, the fact that science and technology work for people, is true by virtue of people working for (adapting to) science and technology. The fulfillment of this requirement, however, can be of such societal or personal impact, that aversion and dissent arise. "Nuclear power is a safe and controllable form of energy" can be a scientifically true statement, but only if this truth is brought about *actively* by securing and controlling the world around it to such an extent that some people find objectionable or unrealistic; to them, "nuclear power is not safe" is therefore true.[52]

In medical science and technology in particular, these two sides of the coin are directly and intimately felt by most people (who in this context, as a first adaptation, immediately find themselves transformed into "patients"). Medicine can help people, but only by virtue of "patient compliance," by virtue of people subjecting themselves to medicine's ways of transforming their problems (with all available instruments and techniques) into treatable diagnoses. In most cases, this price is gladly paid, in others however it is deemed too high, and people start complaining about unwarranted "medicalization"[53] of their problems, of feeling reduced to objects or mere 'cases' in medicine's hands, about medicine's dehumanizing treatments, of its losing sight of the "whole person" and the "real problem".

The point is, however, that a bottomline definition of what the "real problem" is in a given case does not exist. There are only optimal or privileged localizations (through transformations) of problems. No one complains about medicalization when, on getting hurt in a game of football, x-rays reveal a broken leg, and consequent medical treatments take place, though it may be that the 'real' issue is, that one is habitually overestimating oneself and playing far too rough. When a boy keeps getting the wrong answers to questions put on the blackboard in front of the classroom, the first problem presenting itself is that he is doing worse in arithmetic than his classmates; perhaps he is a bit dim, something which might be above all a concern for his parents. Then again, it may be that he is incapable of reading the blackboard correctly from the back of the classroom. He may then be seated right in front of it and the problem dissappears; or he may have his eyes tested by an ophthalmologist. If he ends up with glasses, nobody will question this 'medicalization', although it is hard to say which in this chain of reformulations of the problem should count as 'the real' one, the first one, cause, or effect. Medicine comes in when the symptoms are located in the *individual body*. This is medicine's particular *forte* because at this level it often succeeds in solving problems. That we perceive the medical, or physical problem-definition as the cause, and the other definitions as the mere effects, has to do with cultural habits that systematically privilege certain localizations. Such privileging has many causes and reasons, an obvious one being that, after centuries of cultural investments in the broadest sense of the word, by now it often works rather well in bettering the situation. If *truth* has anything to do with this, it is in the pragmatist sense of the word - 'truth' as that which works. In addition, there is a persistent cultural belief that the objects of medicine, biology, chemistry, and, ultimately physics, however abstract and removed from everyday experience their constructs might be, are more basic, more real and more consequential than anything else in the world. When it comes to medicine and biology, this in turn may have to do with the fact that by now[54] it is often easier to manipulate bodies, rather than, say, social processes, environmental factors, moral convictions, or wherever else certain problems may be localized alternatively. Thus, in some cases, and in some parts of the world, it is actually easier to have the shape of one's nose or breasts changed surgically, rather than change social environments that ridicule certain body shapes, or one's psychological strength to withstand such pressures. So far, it has proven easier to give thousands of children medication for asthma and other respiratory difficulties, rather than diminish the air pollution in my home town; for a long time it was easier to give medication for insomnia, depression and nervousness than to acknowledge the need of housewives to find some fulfillment and self-esteem outside the home.

It can work the other way around as well, with some groups of people having what they experience as physical problems systematically dismissed as psychological ones – ironically enough, infertility used to be an example of this.[55] Sometimes, the very issue of whether or not to define a problem in medical terms is the point of contention in enduring societal and political fights. For some people, unwanted pregnancy will always be a moral one, where others keep struggling to have it defined as a problem of lacking access to safe medical treatment and contraception technology.

This book, however, is not directly concerned with the question how certain problems came to be perceived as predominantly medical ones, calling for medical-technical solutions. It will not address the question of why and how involuntary childlessness is predominantly treated as infertility, that, as a sort of disease, is at home in clinics. Although it may be quite relevant to ask why and how involuntary childlessness has been variously seen as a psychological, social, or even a moral problem, or as simply a matter of god's will or fate, I will start from problem-definitions that are already medical. So too will I pass over questioning why deviations, handicaps, and diseases some children are born with, are not conceptualized in terms of normalization processes and intolerance in society, poverty or lifestyle, cuts in budgets for provisions to care for these children and support their parents, and so on. Though it is possible to do otherwise, I take medical definitions of congenital disease and anomaly as starting points (which, it should be noted, is not the same as their definition as reproductive problems). My focus is not on medicalization per se, but on specific transformations of problems *within* medicine.

4. CONTEMPORARY TECHNOLOGICAL PRACTICES

The second section of this chapter described how the female reproductive body, throughout the history of modern medicine, has been viewed as inherently imperfect and naturally and essentially problematic, a view expressed in its being overexposed to scientific investigation and experimentation, theorizing and intervention. The female body thus became configured as a particularly 'privileged' site to localize problems, a pattern that appears to become self-perpetuating.

This study is about the way the pattern of convergence of ever more problems in the female body currently expresses itself in transformations of men's fertility problems and children's congenital diseases. It deals with two technologies in particular: fetal surgery and the treatment of male infertility through IVF and related techniques.

4.1 In Vitro Fertilization

Since its introduction in the late 1970s, in vitro fertilization has enjoyed wide media coverage and has never been far from the headlines, but few people are probably aware of the actual extent of this practice, its efficiency or its hazards. The public perception of IVF has changed from initial shock and worry into its wide acceptance as one fairly routine, efficient infertility treatment among others.[56] Although this change involved many controversies, these debates, in focusing on often marginal aspects or derivative problems (such as kloning, single or postmenopausal motherhood, sperm-donor anonimity, and so on), actually played a significant role in creating a core definition of this technology as an acceptable, non-controversial treatment with a specific set of unchallenged and legitimate applications. By now, this technology has grown into a world-wide practice.

To give an impression of the extent of the practice some numbers are in order. The latest national survey on assisted reproductive technology (ART) in the United States [57] reported over 65,000 treatments in 1996, more than 44,000 of which were in vitro fertilization treatments (including 14,000 ICSI procedures[58]). The rest consisted of Gamete Intrafallopian Transfers (GIFT[59]), Zygote Intrafallopian Transfers (ZIFT[60]) or some combined treatment. These treatments were provided by some 300 ART programs, a number estimated to be close to the actual total number of clinics providing these services. In the Netherlands, a total of nearly 10,000 treatments was registered to have taken place in 1994.[61] A comparable survey for France[62], covering the period 1986 to 1990, reports a yearly growth of treatments, with over nearly 20,000 treatment cycles in 1990, some 18,500 of which involved

IVF. These numbers were estimated to cover 80% of the total activity in France.

Of the 44,000 initiated treatments in the American report, 22% resulted in the delivery of a child, that is, some 10,000 births. For the Netherlands, the nearly 10,000 treatments yielded some 1,500 ongoing pregnancies, leading to the birth of 1,650 children, which amounted to a successrate of 15%. For the French clinics, this number was reported as 13.6%, but this latter number was calculated per oocyte retrieval, not per initiated cycle. Since a part of the initiated treatment cycles are usually cancelled before oocyte retrieval, the birth rate per initiated cycle is actually still somewhat lower. Subtleties of the latter kind account for the enormous difficulty in finding general and comparable figures for the various aspects of artificial reproductive technologies (ART), as well as for the vast differences between numbers cited in various contexts. Besides huge differences in success between clinics and between the various medical indications for treatment, success rates reported may refer to accumulated birth rates after a number of treatments, or they may refer to pregnancy rates (rather than live births) per embryo transfer or fertilization rate (rather than per initiated cycle), and so on. Thus it is not uncommon to encounter a successrate for IVF of 30-40 % in popular media, whereas the figure of 10-15% also circulates widely. The former figure usually refers to the mean result of the accumulated treatments a woman may receive, which is often three or more. The much lower figure of 10-15% refers to a much more strictly calculated outcome per initiated treatment cycle. Unsurprisingly, advocates usually cite the first figure, while critics insist on the second. To give an impression of why it is important to be precise in recounting the exact reference of the outcomes related in any story about IVF, let me walk through the American figures step by step.

Of the 44,000 initiated cycles (ovarian stimulation protocols started) only 38,000 actually resulted in egg retrievals, accounting for a drop out of 14% in this first phase already. Obviously, the actual attempt at achieving a fertilization in vitro can be done only in cases where eggs have been obtained. The reason for cancellation of an already started hormone therapy may be that the person in question does not respond enough to the medication, but it may also be that she is overresponding: treatment is cancelled of those women who, in reaction to medication, become at high risk for severe ovarian hyperstimulation syndrome (OHSS). This syndrome is a feared, potentially life-threatening complication that in mild to moderate forms is estimated to occur in 5-20% of all hormonal stimulation treatments, and in the severe cases (about 1% of all initiated cycles, justifying an estimation for this particular sample of 440 such cases) requires immediate hospitalization, intensive care, and often surgery to prevent it from becoming lethal.[63]

Subsequently, the number of successful fertilizations in vitro, expressed in numbers of embryo-transfers (ET's), is 36,000. Thus, in the laboratory, another 2,000 of the initial 44,000 treatments are lost. The biggest drop-out, however, occurs in the next step: the real bottleneck in IVF is the establishment of pregnancies after embryo transfer. Only some 12,000 of the 36,000 ET's resulted in pregnancy. Here interpretative caution is still needed, for 'pregnancies established' is not in all cases interchangeable with socalled 'ongoing pregnancies', let alone 'deliveries', or 'delivery of a healthy child': usually, ectopic as well as aborting pregnancies are still included in this figure. Some 1,930 pregnancies aborted spontaneously, and some 50 were ectopic (a dangerous condition requiring quick surgical intervention and, if not detected in time, potentially causing severe damage and lasting infertility). This leaves 10,000 pregnancies progressing toward delivery. Before counting this as the number of happy endings, it most be born in mind that only 61% of these pregnancies were singleton pregnancies, that is, the rest involved twins and triplets and more. Such pregnancies, taking up 39% of the total, are high risk pregnancies, occasionally calling for such painful and dangerous interventions as "fetal reduction" (abortion of a number of the multiple fetuses), and otherwise associated with extroardinary high rates of preterm labor and prematurity, Cesarean sections, neonatal and maternal morbidity.[64] Finally, a number of 202 stillbirths has to be substracted and of the babies born alive, 164 had structural or functional birth defects and 154 died during the neonatal period.

Thus, counterpart to any clean percentage of 'success', however calculated and valuable to the lucky ones concerned, is a multitude of disappointments, tragedies, health risks and actual physical and emotional damage. Although childbearing without medical assistance carries risk and potential for tragedy too, IVF and associated techniques bring along substantial amounts of sickness, pain and grief that should be considered iatrogenic in nature. The ovarian hyperstimulaton syndrome, the elevated numbers of ectopic pregnancies and spontaneous abortions, and the morbidity and perinatal mortality resulting from the disproportionate numbers of multiple pregnancies are the clearest examples. So, quite in contrast to the general public image of IVF, as well as to the general tone in the scientific literature, and suggestions to the contrary implied in the very scale on which the techniques are practiced world-wide, there are some sobering facts about ART. Nevertheless, it is quite rare to encounter an explicit acknowledgment of these facts in the medical literature, as in the following quote.

Since the birth of Louise Brown in 1978, IVF has undergone a considerable development: tens of thousands attempts are made every year throughout the world, resulting in the birth of thousands of children. This trend, added to the fact that the average succes rate remains fairly low (approximately 10% of births per attempt) clearly raises two different kinds of problem: one concerning public health in general and the other an analysis of the method and the factors linked with success.

In vitro fertilization as a method of procreation involves a high financial cost for both the parents and the healthcare system. It is also costly in medical terms for the patients; it is an invasive procedure, and complications are not uncommon, even when dealing with [previously] perfectly healthy subjects. Some form of evaluating the method is therefore required. Second, because success rates are so low, any improvement in these rates can only be demonstrated on large numbers of subjects ....[65]

Against this background, the role of male infertility in this practice yields another set of contrasts and paradoxes. Although IVF attempts with male infertility give the worst results of all cases (the ones with donor semen, whether or not there was a female tubal problem, yielding the best), there has been already for years a steady proportional shift from treatments indicated for female fertility problems towards male indications.[66] In the Netherlands, semen abnormalities are diagnosed as the cause for infertility nearly five times more often than tubapathology, the 'original' indication for IVF.[67] According to figures published by a special workshop group of the European Society for Human Reproduction (ESHRE) on male sterility and subfertility, there is a male abnormality in some 50% of infertile couples[68]. Defining male abnormality, however, constitutes a problem in itself. For instance, included in the figure of 50% normal males in the ESHRE report are men with various abnormalities relating to fertility but whose semen is considered 'normal'. Specific male diagnoses counted only as diagnosis if in addition the semen was abnormal as well. There is an almost exclusive focus on semen quality in the definition of male normality, although, paradoxically enough, both conventional and more sophisticated techniques of semen assessment are generally acknowledged to bear little relation to actual fertility and to have little prognostic value regarding the chances of establishing a pregnancy. According to the same report: "Conventional semen analysis gives poor prognostic information about male infertility. The newer, highly technical procedures have also been disappointing in terms of prediciting pregnancy and in addition are of limited applicability from a practical point of view. At present, no systematic quality control methods have been developed for the various tests."[69] A survey of departments of obstetrics and gynaecology in Western Europe yielded the conclusion that despite the equal distribution of infertility between males and females, "both general and specific examinations were applied more frequently in the female than in the male partner". Also, the criteria for normal semen varied widely among both departments and countries, leading the authors to conclude that "fertility investigations are based more on tradition and personal preferences than on the demonstrated utility of its components."[70]

During the years covered by this study, the most significant developments in IVF involve those regarding its application for male infertility. In the early 1990s several new techniques to be used in the IVF laboratory in order to address male fertility problems, existed alongside each other, each in its first experimental stage; by the time this study is about to be finished, a consensus has been reached by the specialists involved that only one of these techniques actually proved worthwhile.[71] In the mean time, this latter technique, Intracytoplasmic Sperm Injection (ICSI, a term that itself stabilized only after a period of wide diversity) has been introduced in many clinical programs world-wide. While the first reported pregnancies by this technique date from 1992[72], the 1995 volume of "Human Reproduction" listed as many entries on "ICSI" as on "IVF" itself, as an indication of the fast growth in relative scientific interest in this technology. Clinically, its importance has grown explosively as well. The "ESHRE Task Force" on ICSI reported on a total of nearly 24,000 ICSI treatment cycles performed world wide that same year by 101 treatment centers.[73] While up to 1995 some 1,500 ICSI treatments had been performed in the Netherlands[74], over 3,000 were done in 1997 alone, accounting for nearly 30% of all IVF treatments.[75]

The technique consists of the selection and injection of one single spermcell directly into the egg in the petri-dish, as opposed to the 'conventional' insemination involved in IVF, where hundreds of thousands of sperm are simply joined with each egg in the petri-dish. The other steps of the IVF procedure, the stimulation of the ovaries, egg retrievals, embryo-transfers and so on, remain the same. Precisely this selection of one single spermcell is both the strength and the weakness of this technique. It is its strength because with only one cell needed, practically any infertile man becomes eligible for the program, no matter how bad his semen characteristics are. It accounts for its controversiality as well, since the selection "by hand" cuts out a natural selection process that is quite possibly involved in securing healthy offspring: it may be that genetically inferior spermcells are used. Uncertainties and risks like this exist in exacerbated form when ICSI is performed with sperm obtained via the techniques MESA (microsurgical epididymal sperm aspiration) and TESE (Testicular sperm extraction), that is, retrieval of sperm from the testis or the epididiymis by puncture. These techniques are used with men who do not ejaculate or who do not have any spermcells in their ejaculate. While advocated and applied by many practitioners, others have raised doubts concerning the genetic effects of such practices. There are, for example, specific indications that cystic fybrosis and some hereditary forms of infertility are reproduced through the use of such sperm.[76] Moreover, even if it is acknowledged that ICSI treatments result in relatively more children with chromosomal abnormalities, an interesting kind of reasoning still applies:

> The incidence of de-novo chromosomal aberrations in children born after ICSI is slightly higher than expected in the general population, but this fact is probably linked directly to the characteristics of the infertile men treated rather than to the ICSI procedure itself.[77]

The answer to such dangers is generally sought in another "technical fix" in the form of genetic screening, pre-implantation diagnosis and (invasive) prenatal diagnosis[78] (besides, of course, intense scientific research and experimentation made possible by te extensive worldwide clinical application of ICSI). It is even anticipated that ICSI generates its own future clientele: "it is possible that the sons of these infertile couples will also require ICSI when they grow up and wish to have a family"[79] (Note that the reference here is to "the sons" requiring ICSI, while in fact it will be the sons' future female partners who are thus destined in advance to undergo fertility treatment.[80]) For some, however, these risks are considered serious enough to temporarily cancel their MESA and TESA programs,[81] but not the ICSI programs themselves, although conclusive evidence concerning the innocence of this technique is still lacking. Notwithstanding this, several teams already consider the possibility of performing ICSI in all in vitro fertilizations.[82] It appears that this technique is used extensively, once again, in order to *eventually* establish its safety or dangers.

4.2 Fetal Surgery

Fetal surgery, in contrast to IVF, has not gained such wide public or professional acceptance yet, and many doubt that it ever will. Compared to the frequency in which IVF is almost routinely performed around the world, fetal surgery remains a matter of incidents. Instead of the tens of thousands of yearly IVF-treatments on national scales, fetal surgery involves numbers closer to tens and hundreds yearly, performed in a handful of centers. These specialized centers have nevertheless been offering programs for years now, and in their experience they have developed some level of routinization, standardization and improvement of techniques by treating series of pregnant women. Practitioners in this field,[83] though relatively few, have formed their own society, with journals, conferences and concomitant consensus-building.

When addressing medical professionals outside their own circles, however, a defensive tone predominates. As a practice, it is couched uncomfortably between the unstable and sometimes highly charged and controversial "neighboring" practices involving medical abortions, the development of transplantation technology using fetal materials,[84] and neonatal intensive care.[85] Moreover, the indications for surgical intervention are produced by prenatal diagnostic techniques developing more or less at the same time. This means that the interpretation of prenatally discovered anomalies is still rather uncertain, and the implications of these diagnoses for the actual state of the baby at birth usually hard to predict. Prenatal diagnosis itself is such a novel practice that the "natural course" of a particular anomaly seen on ultrasound, that is, its development without intervention, is something that is more often than not still unknown. Some seem to resolve over time; others have done so much damage already that even with intervention the child would not be viable. Moreover, with the pace at which neonatal medicine is developing at the same time, the criteria for postnatal treatability are constantly shifting as well, which further complicates any assessment of the value of experimental and risky prenatal interventions.

When in 1982 the International Fetal Surgery Society (IFMSS[86]) was founded, these problems and limitations were, to some extent, already recognized. Tentative guidelines were formulated for the selection of potential cases, requiring that there be only a single, structural anatomical defect hindering normal organ development, that could be reversed by the intervention. In addition, this defect in all likelihood had to be incompatible with postnatal life or seriously life-threatening[87]. This excluded most prenatally detectable conditions from consideration for intervention, since usually there are multiple anomalies, and in many other cases effective postnatal treatments are available.

It was more or less agreed that three clusters of problems met the requirements of these guidelines: urinary tract malformations or obstructive uropathy, diaphragmatic hernias, and hydrocephalus. Most of the procedures performed during 1980's concerned one of these three diagnoses, but occasionally treatments for others problems were tried as well. However, after the publication in 1986 of the results of procedures for obstructive uropathy and hydrocephalus, gathered by the Fetal Surgery Registry[88], controversy and diffidence about the feasibility of these procedures was stirred up rather than abated.

The results of the procedures for hydrocephalus, acknowledged in the Registry's report to be "not encouraging", even led to an unofficial moratorium. In hydrocephalus an excessive build up of cerebral fluid causes enlargement of the ventricles and elevated pressure in the cranium, damaging the development of the brain. Prenatal intervention for this condition resembles the postnatal treatment: through (serial) puncture or (lasting) catheterization ("shunting"), the pressure in the brain is relieved. However, whether this results in restoration of normal brain development is highly uncertain. Moreover the variation in severity of brain impairment at birth in untreated cases is extreme, ranging from anencephaly to normal brain function.

In the prenatal experiments up to 1985, more than 10% of the fetuses died as a direct consequence of the procedure,[89] while survivors were in large part moderately to severely neurologically handicapped. Significantly, animal studies conducted in the same period as these experiments in humans also show many additional complications, including lethal fetal brain infections.[90] The IFMSS did not officially subscribe to the moratorium[91], but clinical experiments nevertheless more or less stopped, and the special artefact used for the procedure, the "shunt", was taken out of production.[92] While some researchers argued for attempting to improve the results by more drastic interventions,[93] the experience with hydrocephalus has since been referred to as a negative example of in utero therapy, and a warning against overoptimistic experimentation.

Although the results of procedures relating to obstructive uropathy were somewhat less devastating, they too stirred polemics. The problem concerns a blockage in the urinary tract, causing an accumulation of urine, damaging to the kidneys (hydronephrosis). The lack of fluid excretion further causes a shortage of amniotic fluid (oligohydramnios), which has an impairing effect on lung development. Babies born with this problem often die of respiratory problems even before the uropathy is treated (repairing the obstruction, dialysis or kidney transplantation).[94] Prenatal intervention would consist of procedures comparable to the ones in hydrocephalus: puncturing the bladder or kidney, or the placement of an permanent catheter or "shunt", thus draining the accumulated urine into the amniotic fluid. Mostly this is done percutaneously using a hollow needle guided by ultrasound, but "open procedures", i.e. surgically opening the womb and exposing the fetus were tried as well.[95]

Here, the Registry reports large numbers of procedure-related and neonatal deaths, elective abortions for additional (chromosomal) abnormalities and irreversibly dysfunctional kidneys undiagnosed at the time of intervention (totalling 42 of 72 cases). It warns that even these results may be biased towards the more positive cases, since the registration is voluntary, and that the positive outcomes cannot be attributed to the interventions for lack of controls. These qualifying remarks notwithstanding, the report is seriously criticized for its failure to report adequately on complications and failed procedures. A critical review of the same and other cases mentions infections, inflammations, dislodged and clogged "shunts", the consequent need for numerous attempts in 92% of the cases, and premature deliveries triggered by the procedure (mostly resulting in neonatal death)[96]. These problems do not improve over the years to follow, leading many to plead for cautiousness and conservative treatment, that is, no intervention during pregnancy[97]. Others, however, see these problems as inherent to any novel treatment and learning process, and seek improvement through more drastic intervention with more technology[98].

With respect to the third problem initially selected for prenatal surgery, congenital diaphragmatic hernia (CDH), the story was equally dampening. While the 1986 report mentions only one, unsuccessful case in humans, animal studies in this period are soon followed by experiments in humans. In CDH, there is a hole in the diaphragm, causing organs and intestines to "herniate" from the abdominal cavity into the thorax, where they impair lung development (pulmonary hypoplasia). As in the case of uropathy, postnatal treatment is available, but the period needed to restore development of the lungs is often too long to be bridged by the artifical means available in neonatal intensive care. Prenatal surgery for this problem, repositioning of the migrated organs and closing the diaphragmatic hole, is extensive and imaginably invasive, always calling for open procedures (hysterotomy). Many complications arose and many technical variations were tried[99], but when the leading team in this particular area published an overview of its results over 1981-1991 they were admitted to be "frustrating". Of 61 referred cases, 14 were selected for prenatal surgical repair. Of these 14 cases, 5 resulted in fetal death during the operation, three within 48 hours after surgery, and 2 after a prematurely triggered delivery. Only 4 survived, but these were premature and sick. Despite these outcomes, however, the tone remains optimistic: the problems are 'technical', called 'challenges' to be overcome, and to be learned from.[100] The team will go on trying numerous technical innovations, despite little improvement in results.

After more than a decade of experimental, more and less "heroic" operations, the initial sense of medical revolution has sobered significantly.[101] Still, fetal surgery is often described as very exciting and as holding great potential for the future.[102] At present, the risky, open procedures are mostly avoided, and efforts are directed toward the development of somewhat less invasive types of intervention using endoscopy.[103] But although in this type of intervention the womb itself is not cut open, the abdomen is, and the exposed uterus entered with several instruments, which still carries great risk.[104]

A second direction for expectant glances towards the future, anticipating the development of endoscopic intervention methods, is the development of fetal gene therapy. Although the first three, hardly successful in vivo attempts in sheep and mice were reported only in 1995,[105] great potential benefit for humans is expected. Cystic fibrosis, Duchenne muscular dystrophy, some neurological diseases as well as non-inherited diseases like neonatal respiratory distress syndrome, and infectious diseases such as AIDS, hepatitis B, measles, and toxoplasmosis have been mentioned as possible candidates for fetal gene therapy.[106]

During the first years of experimental fetal surgery little systematic research investigated the effects and risks for women (or the pregnant animals used, for that matter). The first publication addressing this particular issue dates from 1986, and concerns a retrospective analysis of experiments on pregnant monkeys. It concludes that "serious maternal complications occurred, including 3 maternal deaths, 5 uterine ruptures, and 5 cases of wound infection" (in a total of 102 procedures); "prenatal intervention carries significant maternal risk."[107] With the series of procedures reported by the Registry covering the period before 1985, this implies that many treatments of humans were executed before any serious attention was paid to this issue even for animals. Retrospective analysis of "maternal outcome" in humans are first published in the early 1990's.[108] Although many authors state that "obviously, maternal safety is the first priority in all cases",[109] or words to similar effect, the timing of the attention to this issue, both in relation to animal studies and to scientific publications, appears to tell a somewhat different story. As does the following remark: "Prior to application of fetal surgery for correction of human malformations, it is essential to document success in the rigorous nonhuman primate model. ... we have demonstrated the feasibility of operating on fetal monkeys without significantly increasing fetal morbidity or mortality as compared with a nonoperated control group. However, the crucial issue of maternal safety with fetal surgical procedures remains unresolved."[110] This would seem to suggest that technical feasibility and fetal outcome literally as well as practically actually come prior to maternal safety, since these results did not postpone clinical application at all.

Meanwhile, the intra-operative complications suffered by the women through these procedures included, for instance, excessive hemorrhaging, hypotension, and respiratory failure.[111] Post-operatively, problems like amniotic fluid leaks (necessitating a second operation), chorioamnionitis (potentially requiring an emergency Cesarean section), enterocolitis, miscarriage, premature rupture of membranes, continuous uterine contractions, necessitating aggressive treatment with tocolytic drugs (indomethacin, terbutaline, nitroglycerine), all of which cause potentially severe morbidity, as well as serious risks for the fetus (like brain injury). In the majority of cases in which the fetus survives long enough, the pregnancy ends in a premature delivery through Cesarean section. Maternal morbidity, if reported as a special category, however, does not include the inevitable post-operative malaise and necessary recovery associated with the fetal operation itself nor that following the Cesarean sections that are required because of the procedure. Also, the risks and complications of the Cesarean sections (in the pregnancy concerned as well as in later ones[112]) are not considered as risks and complications of the fetal procedures, although the first are a direct consequence of the latter.[113] Specifically in the United States, where Cesarean sections are considered quite routine,[114] the attitude toward this form of major surgery sometimes seems to be carried over to fetal surgery as well. Remarks like "While the fetal operation is similar to a Cesarean section for the mother, it is a major physiological stress for the fetus"[115], show how the comparison almost functions to trivialize the significance of the procedures for women.

In general, there is a fundamental tension between the proclaimed priority of 'maternal safety' and the very concept of fetal surgery. This tension is clearly underlying the apparent relief expressed in this quote from two leading experts in the field: "Obviously, maternal safety is the first priority in all cases. *Fortunately*, there has been no maternal mortality."[116]

5. TECHNOLOGIES AS DISCURSIVE PRACTICES

New and contemporary as the technological practices of IVF and fetal surgery may be, they fit the historical patterns of convergence of problems in the female body and uneven distribution of experimental risk rather well. Using IVF on women in order to address male fertility problems should be considered an extreme case of the pattern in which men's reproductive bodies remain unproblematized and untouched by medical science's experiments and interventions, while constituting a strong intensification of such involvement with female reproductive bodies. Moreover, once again, the lack of knowledge and methods for males themselves, and the abundance of it concerning females, is first the *reason* for these practices, but also its probable *result*. Thus a reproduction and intensification of this assymmetry is to be expected.

Similarly, fetal surgery forms the next step on a path that already led to the well-known extensive medicalization of the female reproductive body through medicine's intense problematizing, intervening in, and experimenting with pregnancy and childbirth. It would never have been contemplated without the knowledge produced by intense medical exploration and surveillance of millions of *un*problematic pregnancies and births, and without the instruments, techniques and organizational structures developed to do that. By bringing all pregnancies as potential hazards within medicine's reign, and routinizing prenatal testing, the number of occasions for drawing up diagnoses, discovering new complications and, trying out new techniques has grown explosively, in the end creating a niche where actual surgical intervention came to look sensible.

However, the two technologies are not mere repetitions of such patterns, but rather form new chapters of a history in which new boundaries are crossed. We now find problems being addressed through the female body that are explicitly acknowledged as being the problem of *other* bodies. That is, it is no longer a precondition that the female body be pathologized in order to argue a need for intervention, but actually "perfectly healthy subjects" become included in the pattern. So firmly entrenched are the patterns designating the female body a natural object of intervention, that it has come to seem medically and biologically inevitable to treat others via this route. Thus, these new technological practices extend and reinforce these patterns to considerable extent.

In tracking the transformation processes involved in changing congenital disease and male fertility into problems for female bodies, much can be learned from scientific papers authored by clinicians working in the fields concerned, and involved in medical research. The scientific paper has been identified by students of science of many hues as a rich source for investigation. At a basic level, such papers constitute reports of work that has been done at actual research sites, in clinics, laboratories, or both. As such they provide information, usually in a very economic and standardized style, about what has been done: which problem was studied, how one went about it, what results were produced and how these results should be interpreted. From this angle, scientific papers are seen as neutral carriers of information about work done elsewhere, enabling others, especially colleagues, to get to know this work. Though these papers also function as internal professional communication, this view, roughly coinciding with science's own view on its writing practices, is far too restricted.

Because scientific writing is generally regarded as mere formalized reporting, drawn up after the real work is done, it has been argued that for students of science interested in the philosophical, sociological, cultural and political aspects of science, it contains little of interest: the real work being done in the laboratory, one should study what 'really' happens there, as opposed to its post-hoc representation in published reports. However, this is a rather limited view of what such papers are and do. The scientific paper should be considered as an important part of the work itself, rather than a mere reflection of that work, the adequacy of which may then be questioned. As argued, for instance, by Bruno Latour:

> Sientific texts, to be sure, have no privilege, but neither are they inferior to the many sources we have for understanding science. Indeed, when properly studied, they offer a convenient model to show how many mediations can be retrieved from the scientist's own practice. A scientific text is not only a more or less transparent medium to convey information to the author's scientific colleagues, nor is it only a document to help historians, psychologists, or sociologists retrieve the state of mind of its author or the context in which it has been written. As many decades of literary theory have helped us to see, texts are a little bit less and a good deal more than information and document. They build a world of their own that can be studied as such in relative and provisional isolation from the other aspects. They are localized events, with their own matter and their own practice.[117]

Since there is no need to oppose texts to practice (writing being a form of practice), language to reality (language being co-constitutive of reality), there is no need to privilege one particular research site over another, as long as one is aware how the site of choice affects what one is able to see, and provided there is enough of interest there to be seen. In this latter respect, it is precisely the multi-faceted nature of scientific papers that renders them so interesting, as well as amenable to a variety of analytical perspectives and questions. There are many more levels on which the scientific paper operates, many more interesting views on what language does, than the report-of-practice or representation-of-reality views allow. It is precisely by virtue of their multi-faceted agency and functionality that such texts should be considered key elements in the development of science and technology, and hence in the shaping of our societal practices. To clarify these characteristics, let me elaborate some of these aspects.

What makes the scientific paper stand out from other types of texts or verbal utterances is its status and role in the establishment of scientific facts. No finding, however interesting or pathbreaking, exists officially, until it is published in a recognized scientific journal. Scientific literature is something very close to an official record of established scientific truth. It constitutes what David Locke calls "archival writing, intended to be part of the ongoing written documentation that traces the course of science as it proceeds. The scientific utterance is not a casual remark; it is a studied statement, part of the ever growing archive of science. What science consists of, to repeat the figure, is stratum after stratum of written documents recording the continually changing status of current science. The documents of science encompass the progress of science; they *are* science; what science is is what scientific documents say."[118]

A system of peer review and editorial selection and policy serves as a gatekeeper for what can count as genuine knowledge, genuine fact in a particular field. Although even after this point the fate of a particular claim is still open to dispute, refutation or reinterpretation by subsequent readers,[119] it is then on a public record that confers status on its claims and contentions as scientific finding. In some instances it may be ignored by the relevant community, in which case it is forgotten and prevented from becoming an established fact, but in others it may be taken up, cited and referred to by the next author, thus strengthening its status as fact and its function as building block for future developments. In Western science-oriented societies, there is little that comes closer to the status of being the rock bottom of undisputable truth, knowledge and fact than this public, published record of scientific findings. Whatever claim in whatever area one wants to put forward, it becomes stronger to the extent that it is 'backed up' by references to scientific publications. This literature has a cultural authority in deciding what is true that cannot be questioned but by the smallest elite of close colleagues and recognized experts.

Another function of the scientific paper - suggesting its extreme cultural, social and political significance - is that it is one of the most important forms in which facts and technologies leave their places of origin to be dispersed through the world, changing it along the way. 'Spread of information' hardly captures the very material and practical sense in which scientific practices and technologies spread through society, not exclusively but certainly partly through the mobile text[120]. In this respect scientific texts are perhaps better conceived of as constituting a manual, recipe, or script[121], enabling the dissemination and standardization of the concepts, questions, techniques, conditions and procedures needed to reproduce the knowledge or technology concerned elsewhere. Thus, instead of being mere *post hoc* representations of what has been done in one particular research site, they are part of the necessary building blocks, the *constitutive* elements of similar work and practices elsewhere. As John Law phrased it, scientific articles give the laboratory "the capacity to act at a distance upon the world in all its diversity. ... It is the text most of all that the laboratory uses to rebuild the world. It is the text that boxes in and regulates the points of contact between clinicians and researchers, patients and fund-raisers, laboratories and diseases. In short, it is first and foremost the text which imposes a structure on the world."[122]

In the case of the medical-scientific paper, there is an additional aspect, touching upon the definition of medicine as a science. Medicine is a practice where doing science and providing healthcare, research and its application, are often thoroughly intertwined. In clinical practice, and especially the clinical trial, the same actions and interventions may constitute both a therapeutic effort on behalf of a patient and a step in an experimental procedure or protocol. They are then distinguishable only on the level of goals that may or may not partly conflict with each other. Moreover, each goal simultaneously and ultimately serves as the rationale or justification for the other. Medicine needs the scientific aspect in order to enhance its status, credibility, and effectivity, whereas the scientific work needs the clinical purpose as justification for its experiments and 'use' of (usually sick and dependent) people. In this narrow and closed circle, the publication of articles is instrumental in creating a somewhat clearer distinction, a post-hoc disentanglement of the two: the papers constitute the output in terms of facts and knowledge extracted from the clinic, thus embodying the scientific aspect voided of immediate clinical concerns. When clinicians have a reputation as scientist as well, they do so by virtue of their publications in scientific journals. In the act of writing and publishing, something is created that leaves the clinic, in a form that endows it with factual, universally valid status, after which detour it then reenters the clinic as scientific input, as premise on which further actions, therapeutic or scientific, are based.

For these reasons such papers more than deserve scrutiny, and from different perspectives than internal professional information gathering alone. As an object of study, the scientific article has proven to be amenable to analysis by a range of methods from diverse disciplines: history of science, literary criticism, rhetoric, discourse analysis, semiotics, and science and technology studies[123]. From such efforts, several insights about the scientific article have come to the fore, that are relevant to the purposes of this study. Among these are the identification of features that give the scientific text its specific power to 'act at a distance' and effect changes in the world beyond its place of origin.

Next to such obvious characteristics as being durable, easily transportable and reproducible, and thus highly diffusible[124], it is above all, of course, the content itself that is of interest. What has been gained from analysis of scientific texts as text, as writing practice, and as language, by methods drawn from literary criticism, is that integral to this content of the scientific paper are elements and aspects that had long been thought to be absent in scientific writing. Thus it has been shown for example, that far from being objective, neutral mediators of facts and results, rhetoric, metaphor and style do play a role in scientific texts. More important even, they cannot be stripped of as superfluous layers, embellishing or contaminating (according to taste) the real, purely scientific content, but are intrinsically tied up with this content. Contrary to the official scientific view that rhetoric is for politics, style for literature, metaphors for poetry, none of which is concerned with objective truth as is science, such analyses have shown how this official view is precisely the *result* of scientific styles, rhetoric, and metaphors.

"It is a hallmark of the official rhetoric of science that it denies its own existence, that it claims to be not a rhetoric but a neutral voice, a transparent medium for the recording of scientific fact without distortion".[125] Thus, for example, the purported objectivity of scientific language (by virtue of which it was thought to be exempt from stylistic analyses) has been shown to be largely the effect of a particular style called "agentless prose."[126] This is a pervasive and very powerful rhetoric that has the effect of de-subjectifying observations and experiences reported. Instead of "I collected and analysed 75 bloodsamples", one writes "75 bloodsamples were collected and analysed". A simple device by which the scientist-writer absents himself from the scene, distracts from his agency in the acts described, thus creating in language the impersonality that is supposed to be a defining characteristic of science itself. Similarly, it has become clear that metaphors and analogies are not invoked to merely explain or elaborate on the "real" scientific concepts and models, but actually are often at the heart of (the most "technical" of) such concepts; models are "the scientist's version of the poet's metaphor".[127] Since analogies and metaphors are, by definition, drawn from other realms of experience, be it another branch of science or a different context altogether, as a way to render the unfamiliar more comprehensible, identification of metaphors and analogies in science constitutes an apt method to chart cultural content in putatively pure science. Van Rijn – van Tongeren (1997) goes even further, and concludes that it is "especially in the sciences, in which according to some people, language plays only a subordinate role, metaphor sometimes turns out to be of fundamental importance."[128] With respect to medical science in particular she states:

> Highlighting and hiding, one of the most fascinating aspects of metaphor, is of special importance where medical theories are concerned, as valuable therapeutical possibilities may be hidden by the metaphors constituting those theories. [...] The direct link between theory and practice, which is unique to medical science, is illustrated by the analysis found in [medical] texts. Theories and therapies based on them are often formulated in the same metaphors. Metaphors thus determine the therapeutical measures doctors and patients live by."[129]

It is these kinds of textual aspects - discursive mechanisms, as I will refer to them - that will be traced in this study in a body of medical scientific texts. My corpus consists of some eighty papers on the subjects of assisted reproduction techniques as treatments for male infertility, and of a variety of experimental prenatal surgical approaches to congenital anomalies. The papers cover a period of 15 years, and appeared in leading professional journals in both fields[130]. This selection can be considered a collection exemplary for the type of discourse generated in both domains of practice.

Furthermore, the papers analyzed cover a period in which the use of the two sets of technology are still points of contention within the medical community. Both IVF as a treatment for male infertility and the treatment of newborns' congenital diseases through prenatal intervention are still new approaches in experimental stages. Therefore, as of yet the new problem-definitions are not so self-evident, that they do not in many cases still need explicit comparison and connecting with more conventional definitions, in order to establish that these new technologies are indeed answers to the same problems that existed beforehand. This makes the conceptual, the rhetorical as well as some of the medical-technical work necessary to achieve these transformations visible in the texts.

The fact that scientific articles form the empirical basis of this study naturally carries implications for the scope of the findings presented. For all their significance in the development of certain scientific or medical fields, such papers, as was discussed above, cannot be considered direct or faithful representations of clinical practices and events. If, for example, a particular paper proposes a certain approach to a clinical problem without mentioning this or that alternative, it cannot be inferred that such alternatives are not discussed with the patient or with a colleague in the clinical setting. Or, if from the analysis of the role of patients in these papers it is concluded that they are construed in a particular fashion, occupy a specific role or position, this should not be taken as a statement directly pertaining to the position and role of patients in clinical settings, as a statement about doctors' actual "bed-side-manner", let alone about how such settings and the events described are experienced by such patients themselves.

So, whatever the analyses in the following chapters bring to the fore, they must be read and interpreted in relationship to what their objects, the scientific texts, are about. They must be seen as discursive mechanisms and patterns operative within scientific literature, so their relevance is defined as narrow and as broad as this textual realm goes. Although this may seem a severe restriction, the discussion of the scientific paper presented above indicates otherwise. Given the power of scientific texts in the development of science and technology, it becomes very important what kind of world these texts are building. What is present or absent from the textual part of scientific and technological practices becomes highly significant. If, for example, this literature tells us little about the experiences of the women patients whose stories (however implicitly, and in however abbreviated, formalized terms) it tells, this fact in itself may teach us something significant about the role or attributed relevance of such experiences in the scientific rendering of problems, in the communications between researchers, in selecting what is eligible as being of objective scientific value, in counting technological success or failure, or in the expansion of certain technological practices beyond their place of origin.

CHAPTER 2

THE MAKING OF THE NEW PATIENTS

1. INTRODUCING: THE COUPLE AND THE FETUS

To speak of fetuses and couples as patients may seem neither particularly surprising or consequential. After all, we all know that pregnant women carry fetuses in their wombs that may have something wrong with them. In such cases, it seems hardly strange to talk of these fetuses as little patients in the womb. Likewise, couples who, despite serious efforts, are unable to have children, have a problem that may lead them to seek medical help together. But everyone knows that fetuses grow inside women's bodies, and it is women who will visit doctors, who ask for and are given advice, prescriptions and tests. Similarly, everyone knows that a couple consists of two individuals with separate bodies. They may have a problem as a couple, but the shared nature of their problem stems from their shared wish for a child and shared grief about its remaining unfulfilled. So, while it seems self-evident that fetuses and couples may have medical problems, to call them 'patients' is just a manner of speaking, not to be taken too literally. At any time it will be clear who the 'real' patients are. For all practical purposes, it would hardly seem to make any difference. However, Meerabeau, drawing on observations of 55 clinic sessions in three fertility clinics in the United Kingdom, concluded: "Doctors are not accustomed to treating more than one patient simultaneously, and the use of the concept 'couple' in subfertility treatment presupposes a commonality of aims which may not exist. ... There are attempts to construct the fertility problem as a joint endeavor, but these tend to founder on the biological imbalance in the situation."[131] Moreover, This view about the innocence and inconsequentiality of language and conceptualizations relies on the idea that the relation between language and reality is one in which language refers to an independent reality. This view implies that changes in vocabulary have no consequence for the reality described, since language only passively "mirrors" this reality. In this work I rely on a theory of language that accords a much more active role to language in the constitution of the realities we inhabit.[132]

In this chapter I will develop the argument that the notions of the couple as patient and the fetus as patient are simultaneously more 'real' *and* stranger than their current prevalence in medical-scientific discursive practices suggests. Fetuses do not just figure as patients in the fancy titles of the many recent textbooks, articles, and reports on current developments in prenatal medicine. The concept of the fetus-as-patient has gained a presence much wider than that. Similarly, the couples in infertility medicine play the role of patient in a much more literal sense than might be expected from the everyday use of the word. Within the discourses on fetal surgery and in vitro fertilization, 'fetuses' and 'couples' have come to occupy positions very similar to those of more conventional types of patients. Like patients in general, they are referred to clinics, undergo tests, and receive diagnoses and treatments. While their patient status has thus become 'real', at the same time this development constitutes a departure from the meanings 'fetuses' and 'couples' have in contexts other than reproductive medicine, as well as from more conventional meanings of what being a patient is. In as much as they have been turned into real patients, 'fetuses' and 'couples' become rather strange entities.

First, we will take a closer look at the 'couples' populating the discourse on male infertility and reproductive technology. The term 'populating' intends to underscore once more that my approach to these texts will primarily be a semiotic one, in the sense that the question what the terms and categories deployed refer to *outside* these texts is (temporarily) bracketed. As Bruno Latour, explaining the value of a semiotic approach to scientific texts in the study of science and technology, writes: "Semiotics is the ethnomethodology of texts. Like ethnomethodology, it helps to replace the analyst's prejudiced and limited vocabulary by the actor's activity at world making."[133] Moreover, the focus is on the world that is emerging from the texts, without reference to the presumed intentions of the author or the social context. Thus, I focus on *internal* referents and meanings generated by the texts themselves.

From this perspective, infertile couples are far less ordinary than their strong resemblance to the well known social category that denotes a combination of two individuals somehow belonging together might suggest.[134] The rapidly developing discourse on infertility treatments seems to have construed a creature that - although called a 'couple', or sometimes a 'male factor patient' or an 'infertility case' - seems more adequately described as a hermaphrodite being, rather than a combination of two recognizable individuals of different sex.

Numerically, for instance, couples are not counted as two patients, but as *one*. Papers reporting clinical research involving only a small patient sample commonly refer to individual cases with a number. A man and a woman who make up a couple are counted, in these reports, as one, so that when there are, for example, 15 couples involved in a study, the total patient count is 15 rather than 30. Naturally, a reader will take "patient no.6" to refer to one particular human being. This patient might be said, for example, to suffer from oligospermia. The reader then will infer that patient no.6 is a man, because she knows that oligospermia is a pathological condition of the male reproductive system, in which the sperm produced contains abnormally few spermatozoa. When next, however, the reader is told that this same patient underwent an embryo transfer (ET) and became pregnant, she starts suspecting that this patient is not exactly an average male human being.

One might think that such phrases are just one author's occasional slip of the pen, resulting in an accidental omission of words like "the wife of", in the sentence about ET and pregnancy. The same phenomenon, however, turns up again and again throughout the literature. Consider the following examples, taken from scientific publications on the use of IVF in cases of male infertility:

> Ooplasmic injection (single sperm heads) was done in 38 oocytes from three patients with extremely severe oligozoospermia; only four pronuclear zygotes were obtained and replaced into two patients, without any resulting pregnancy.[135]

> In severely teratozoospermic patients, significantly fewer partially zona-dissected than subzonally inserted embryo's implanted.[136]

> Fig.3 Ongoing pregnancy rate per cycle in different groups of men with the corresponding lower limit of sperm concentration.[137]

In the first two quotes, we find the same patients being oligozoospermic or teratozoospermic, as well as producing oocytes and having embryo's replaced into their bodies. The third quote shows how far this discourse is removed from most other discursive practices. Without a trace of irony the authors claim the achievement of pregnancies not just in couples, but literally in men. However, instead of ascribing to the authors a rather incredible ignorance about the facts of life, quotes as these are perhaps better interpreted as showing the degree to which couples indeed have come to be considered as one functional organism in this practice. They have become true hermaphrodites: one patient, with both male and female physical characteristics.

The conceptualization of a fetus as a patient is a departure from long standing conventions as well, but some other nuances are involved here. While the conception of the fetus as a patient in itself might have allowed for the woman to retain her conventional status as the patient in prenatal care (thus yielding a "double patient"), this is very rarely the case. Each 'case' yields one patient only, and though it can remain ambiguous for some time in the course of a particular text, whether this 'patient' refers to a woman or a fetus, this ambiguity is usually resolved at some point, as in the following example:

> In their review of 74 fetuses with bilateral hydronephrosis, they reported results on 16 patients who were defined as having good prognosis. Nine of 16 patients had intervention, and 7 of 16 did not have intervention. Of the 9 patients with intervention, 8 were delivered with normal renal function. Of the 7 patients with no intervention, 2 died, and 2 of the 5 who survived have chronic renal failure.[138]

To have a good prognosis, an intervention, and to be delivered are things that could be said of a woman as well as a fetus. Yet the last two sentences resolve the ambiguity, for the renal function, death, and survival of 'the patients' clearly apply exclusively to the fetuses. In another case, the patient "was referred at 23 weeks with anhydramnios, bilateral moderate hydronephrosis/hydroureter, and an enlarged bladder and proximal urethra."[139] Again, the referral and diagnosis of anhydramnios (lack of amniotic fluid) could still indicate a woman patient, but the rest of the diagnosis undercuts this interpretation for it refers unequivocally to the state of the kidneys and the urinary tract of the fetus. Usually, however, there is less ambiguity. In the following examples it is the fetus who is undergoing diagnostic procedures and interventions:

> Ten fetuses had undergone diagnostic catheter placement and in utero renal function testing. This led to placement of a therapeutic indwelling catheter-shunt in seven fetuses (three required multiple shunts) and a suprapubic vesicostomy in another.[140]

> Twenty-two fetuses with bilateral CH [congenital hydronephrosis] underwent either a diagnostic procedure, a therapeutic procedure, or both.[141]

An additional potential problem, therefore, is that those patients with a large volume of liver in the chest may not respond to antenatal therapy, as the lungs may be primarily hypoplastic and incapable of growth when the viscera are removed.

Nevertheless, because of their expected high mortality, these are the very patients on whom it is most easy to justify antenatal intervention.[142]

> ...a second fetus with immunodeficiency disease was treated prenatally in 1989. This second patient was a younger fetus with a

complete form of severe combined immunodeficiency disease. He was
treated with FLT in june 1989, at the age of 26 fertilization weeks,
....[143]

In a similar vein, the results of the procedures and the concomitant
complications are described as pertaining to the fetus exclusively. For
instance, the report from the International Fetal Surgery Registry[144],
purporting to give an overview of the results of all the registered cases of in
utero therapy for obstructive uropathy, hydrocephalus, and diaphragmatic
hernia up to 1985, relates these results exclusively in terms of 'fetal
outcome'. All "cases" reported are of "treated fetuses", and all evaluative
categories (including numbers of stillbirths, neonatal deaths, survivals with
or without handicap, procedure related deaths etc.) refer to the fates of
fetuses and children. Not once are 'women' mentioned throughout the report,
not even in the category of 'complications'.

While a 'couple' can be understood as two patients becoming one,
resulting in a hermaphrodite being, it is less clear how to describe what
happens in the case of fetuses. Although it would create a nice symmetry to
understand this see as a process of one patient becoming two (the double
patient model), the examples given above show that this is not exactly the
case. There still is only one patient. One possible interpretation is to view
this is an instance of "pars pro toto": 'the fetus' as a figure of speech that
names a part to stand for the whole, a pregnant woman. Though perhaps
somewhat impolite, this would render the fetus-as-patient an innocent figure
of speech, leaving the woman's position unaltered.

However, many feminist critics of developments in prenatal medicine and
technology have taken a less sympathetic view on the phenomenon of fetal
patients.[145] They worry that putting the fetus central stage may negatively
affect the position of the woman as the primary focus of medical concern.
Their analyses have focused mostly on prenatal diagnostic technologies,
such as ultrasound and fetoscopy. These techniques produce visual
representations of the fetus that literally remove the woman from the picture.
Her receding into the background, or reduction to 'fetal environment', is
taken as a possible sign of a diminishing relative weight of her interests in
medical considerations. My examples above do seem to support this
conclusion: where the fetus becomes the patient, the woman no longer
appears to be.

But the fact that women are not represented as *the patient* leaves open the
question how they *do* figure in this textual practice. Despite instances like
the Registry's report, it is hardly conceivable that they are not in some way
or other acknowledged to be present. The significance of the semiotic

reading that ultrasound images individualize the fetus, while reducing women to empty background is not easily assessed.[146] I want to postpone such a general conclusion for now, and first try to make more sense of the process in which fetuses are construed as patients. Another branch of reproductive medicine, infertility treatments centering around in vitro fertilization, also shows the construction of a new type of patient apparently replacing 'women', while similarly providing new rationales for medical intervention in female bodies. This fact suggests that it might be worthwhile to compare the two cases.

At this point, however, it does seem warranted to conclude that fetuses and couples are both more real and stranger than their quick public acceptance suggests. They have come to figure widely and in unsuspected ways in medical discourse, while departing in as many ways from conventional notions about what can constitute a 'patient'. The existence of fetuses and couples in medical representational practices concerning reproduction, leaves many questions unanswered, most notably questions about how these patients relate to the conventional individual patients they seem to represent or even replace. Moreover, if they have more reality than mere figures of speech, the first step in understanding their nature should include an exploration of the questions where these realities hold and how, and at what costs, they are sustained.

2. THE PROBLEMATIC ORIGINS OF THE NEW PATIENTS

To understand how couples and fetuses came to be seen as singular patients, it helps to think of them not in abstracto but in as concrete terms as possible. Their existence is intrinsically tied to the specific medical-technological practices, by which they were generated. To a large extent their existence begins and ends in those contexts. One might of course refer to 'couples' or 'fetuses' in other contexts, but these will not possess the same properties and defining characteristics as the ones investigated here. For instance, when inviting guests to a dinner party, one might send invitations to 'couples', thus treating them as units, but one will not actually put up only one chair at the table. The property of physically counting as one is particular to the couple of infertility medicine. Moreover, to say that fetuses and couples have an existence in particular contexts does not merely imply a contrast between 'medical language' and 'real life'. Medical practices are forms of 'real life', as much as organizing a dinner party is. Likewise, fetuses have been around throughout history: what was imagined to be inside a pregnant woman's womb has been represented in images and words for centuries. But the putto-like human figure drawn in sixteenth-century anatomical drawings of pregnant female bodies, resembling a rather fat three years old child, is not the same fetus we encounter in prenatal medicine today. Saying that 'our' fetuses and couples have existences restricted to highly specified medical-technological contexts, however, is not a denial of their reality. If medical practices - and I mean this as including clinical research, technologies, as well as its representational practices - create contexts where couples and fetuses are given the role of patients, and thus for all practical purposes are treated as real, there is no (relevant) sense in which they are not realities, but 'just' figures of speech.

If, then, couples and fetuses are realities generated in particular contexts, then the next question is: what are these contexts? They were obviously construed and defined *in relation to specific medical problems.* Whereas our concept of couples originated in medical technological practices dealing with fertility problems, fetuses came to count as patients where congenital diseases were at issue. Both came into being in the very process of dealing with those problems; they are the results of particular directions taken in medical problemsolving.

Although in therapeutic infertility practices, particularly those involving reproductive technologies, increasingly the couple and not the woman is considered the patient, this habit has consistently been viewed by medical scientists as a matter of therapeutic pragmatics.[147] Commonly, infertility can be traced back to either male or female pathologies, but in a considerable portion of cases there is a problem in both partners.[148] If the cause of infertility is not known yet, it makes most sense to direct diagnostic procedures at both partners rather than one of them. Commonsensical as this may sound, it must be born in mind that, until recently, the search for causes of infertility was geared almost exclusively toward women.[149] Though in most cases it is still the woman who first visits a doctor if pregnancy does not occur, an analysis of the partner's semen is today usually done early in the diagnostic process. The shift from individuals to couples reflects the inclusion of men in the diagnostic process and is generally considered a step forward from what is now viewed as former medical irrationality and near-sightedness. The inclusion of partners of women seeking treatment also provides the opportunity to devote more attention to (psycho-)social and relational aspects of the problem.

Originally, IVF was introduced as a by-pass procedure for blocked fallopian tubes in women. Oocytes were taken out of the body and replaced into the uterus after being fertilized in vitro, thus avoiding the passage through the fallopian tubes. In this context there was no need yet to break with the conventional way of designating the woman as the patient. From its inception in the late 1970s, however, researchers viewed IVF also as a door to a world of new possiblities in both research and therapy. Next to the study of the processes of fertilization and early embryonic development that was made possible by IVF, therapeutic possibilities for forms of infertility not related to tubapathology were considered early on. In addition to extending the range of female forms of infertility, male sub- or infertility was soon included in the list of possible indications.[150] Trying to fertilize an ovum in a dish under a microscope provided an unprecedented window on the behavior of semen as well. This intensified visibility of men in the medical approach to infertility led to the recognition that traditional ways of assessing the quality of semen were indeed primitive and superficial, and poorly correlating with chances of fertilization in the laboratory. While this generated an increased recognition of 'male pathology', it was found that in some cases semen of men slightly subfertile by the conventional clinical standards could prove sufficiently fertile under laboratory conditions. Today, the link between male infertility and IVF is so firmly established that one researcher even called it "archaic to view male factor infertility separate from in vitro fertilzation (IVF) and treatment of the female partner."[151] It is here that the notion of a couple as one patient, as we encounter it today, finds its origin. While still directed at one body, the female, a therapeutic practice was developed to deal with problems that could reside in a different body altogether.[152] In linking IVF to the problem of male infertility, the notion of the couple plays a crucial part.

The notion of the fetus as patient, in the sense of suffering from disease and abnormality and being the object of therapeutical or surgical intervention, is equally new. in contrast to the visibility of male infertility, the problem of congenital disease has a much longer history. Yet the practice of ascribing disease and abnormality to fetuses, instead of newborns, is of quite recent date. Most historical accounts in the medical literature take 1963 to be the starting point of medical interventions on behalf of the fetus, when A New Zealand doctor named Liley reported the first successful blood transfusion in a fetus, undertaken to solve the problem of Rh isoimmunisation.[153] At the time, this anomaly was a problem with a high incidence as well as a high perinatal mortality rate. Therefore, Liley's report was met with great enthusiasm and followed by many similar attempts elsewhere. The persistent high complication-rate, however, and the introduction of other more successful treatments based on medication with Rh immune globuline, quickly reduced the importance of these first invasive treatments.[154] Thus, with the disappearance of the problem, the concept of the fetus as patient of invasive treatments retreated into the background for another fifteen years.

It gained ground again with the proliferation of diagnostic techniques in prenatal care. Since the days of Liley, the increased possibilities for establishing prenatal diagnoses by a variety of techniques, such as ultrasound, amniocentesis, and chorionic villus sampling, has fundamentally changed the medical care of pregnancy. Ultrasound in particular has become a widely and routinely used technique in most Western countries, producing real-time visibility of the fetus in an unprecedented way. These techniques enabled a new approach to problems of congenital disease and abnormality in that they provided prognoses on the state of newborns in a relatively safe way and with revolutionary reliability. A rapidly growing number of afflictions could now be predicted with reasonable accuracy, offering pregnant women a choice whether to give birth to such children or not. In the positive reading of this development, this shifted the significance of the problem of congenital disease, at least for the afflictions detectable, from the realm of fate and tragedy to the realm of choice. However, the controversy concerning these techniques has not abated yet, and the critical view many people take on this issue concerns the question to what extent real choices, that is, free choices, are being offered here. Moreover, the "choice" created continues to be a wrenchingly tragic one, since abortion of a wanted pregnancy was offered as the only "solution" available to the problem.

Even among medical practitioners who enthusiastically promoted widespread adoption of these diagnostic techniques, the way in which they now had to address the problem of congenital disease sometimes created unease and frustration. With abortion as the only possible solution, they sometimes felt themselves to be, in the words of Liley, on a "search and destroy mission", which was a far cry from the professional self-image they preferred to maintain.[155] Thus, in the accounts of many doctors involved in fetal surgery, it was above all the development of prenatal diagnosis that formed both the necessary condition, and the strongest possible impulse to take the step toward conceptualizing the fetus as a patient in need of therapy. After creating the possiblity of diagnosing a problem in utero, what more logical step could there be than trying to treat and intervene in utero?

After Liley's experimental blood transfusions in the 1960s, it was in the early 1980s that new forms of treatment began to be investigated. Early in this development, the specialists experimenting in the new field convened and reached a consensus on what problems were thought amenable to prenatal surgery.[156] In view of the highly experimental and risky nature of the procedures proposed, they agreed to restrict themselves to structural anatomic defects most likely to be incompatible with neonatal survival. Further restrictions included that there be no additional anomalies, e.g. chromosomal ones, and no equal probability of success of postnatal intervention. This excluded most prenatally detectable anomalies, and led to the identification of three sets of congenital problems as possible candidates for surgery: hydrocephalus, hydronephrosis or urinary tract malformations, and diaphragmatic hernias. Though the selection of these three are often presented as following from the rationality of the criteria mentioned, it is no coincidence that two of the three anomalies are characterized by large accumulations of fluids. Fluid filled masses are relatively easy to recognize on ultrasound images[157], underscoring the constitutive role of available diagnostic techniques in what came to count as "prenatally correctable anatomical defect."

Despite this initial restriction, however, and despite the fact that even for these high mortality associated conditions no practice had been established yet that could be judged safe or effective, attempts to broaden the scope of indications for surgery to anomalies that were non-life-threatening, or for which effective postnatal treatment was available, soon followed. Already during the 1980s, and increasingly in the 1990s, experiments have been conducted in treatments of anomalies like hand deformities, cleft lip or palate, various types of immunodeficiency, as well as cardiac and neurological diseases.[158]

It will be clear that both the notions of 'the couple' and 'the fetus' as patients find their origins in contexts where a response to particular sorts of *medical problems* was at stake. The following section will therefore focus on the ways in which these particular problems are being *transformed* in these technological practices.

3. TECHNOLOGIES OF SHIFTING THE PROBLEM

When we remind ourselves of the fact that congenital diseases used to be problems of children, and male infertility a problem of men, we can notice that in recent years something remarkable has occured. Instead of children and men, we now see fetuses and couples suffering from these problems. This change gives rise to many questions. For instance, what exactly does it mean to treat fetuses and couples as patients? How exactly did the attention shift from men and children to couples and fetuses? Why did this shift take place at this particular moment? What is the role of medical technologies in all this? In approaching 'fetuses' and 'couples' as patients, where does that leave 'women'? In short, what has happened to the perception of men's and children's problems, the solution of which now depends on intervention in female bodies? Questions like these are not explained by pointing to the emergence of new types of patients alone. Quite the opposite is the case. To arrive at more interesting answers to these questions, one has to treat fetuses and couples and their patient status as explanandum instead of explanans. If one seeks to understand why today women's bodies are operated upon for problems that used to belong to others, 'fetuses' and 'couples' may, at first glance, look like part of the answer, but their appearance is in fact more adequately viewed as part of the change to be explained. It is part of a larger process that has resulted in a transformation of the problems of male infertility and congenital disease. What exactly happened to these problems is the subject of the next sections. The mechanisms involved in in vitro fertilization for male infertility and fetal surgery for congenital disease are not quite the same, which is why they will be treated separately here.

3.1 Working from Bodies to Laboratories and Back

Earlier I suggested that some of the more interesting questions to ask about the couple as a hermaphrodite patient are how, and at what costs, this construct is made to hold. In order to answer these questions, we have to look first at the way the medical problem is being construed. How did the issue of male infertility become transformed into a problem whose solution has little to do with male bodies, but all the more with female ones? What were the steps taken to bring about this shift in definition and location of the problem as well as its solution? How did male patients leave the scene so easily, and did we get 'couples' in return?[159] In this section several different, but often related transformations are described, each of which contributes to the creation of the logic of treating male infertility with IVF.

The following quotation, the opening paragraph of a scientific paper, is a telling illustration of the sequence of steps characteristic of the kind of transformations involving IVF as a treatment of male infertility. It illuminates how such transformations involve different localizations of the problem, crossing individual, male and female, body boundaries.

> The severe oligospermic patient usually has a combination of multiple sperm defects, with very poor chances of spontaneous conception. De Kretser and co-workers have reported that when there is a combination of three or more defects in the semen analysis, fertilization in vitro diminishes to <8%. Tubal embryo transfer (TET), otherwise known as pronuclear stage transfer and tubal embryo stage transfer, has been thought to improve outcome with male factor infertility. However, it does not offer hope for such patients because the spouses' oocytes still need to be fertilized before transfer into the fallopian tubes.[160]

In these first sentences, the central character is a male patient: the oligospermic patient, that is, someone suffering from an abnormal low concentration of sperm cells in his sperm. It is also suggested that such a patient usually has other sperm defects. Furthermore, he is said to have a spouse, and it is to this male patient the technique may or may not represent hope. In each of these sentences a redefinition and a relocalization of the patient's problem occurs. The following is a reconstruction of the steps by wich the redefinition of the problem proceeds.

1 The patient is oligospermic
2 The sperm shows defects
3 Spontaneous conception is unlikely.
4 Fertilization in vitro is highly problematic
5 Outcome [of IVF] may be improved.
6 Spouses' oocytes need to be fertilized [in vitro].

In this sequence, the location of the problem is changed with each step:

In 1, the problem lies with a male individual;
In 2, the problem is a property of the sperm, a substance secreted from and subsequently existing outside the body;
In 3, the locus is a woman's body as the site where spontaneous conception does   or does not take place;
In 4, the problem occurs in a petri-dish in a laboratory;
In 5, the problem is back in the female body, for the 'outcome' to be improved designates pregnancy or delivery. Significantly, the improvement discussed here, tubal embryo transfer, represents a shift from the uterus (normal ET) to the fallopian tube, both locations in the female body.
In 6, finally, the problem has returned to the petri-dish but, in the process, it has become linked to the oocytes.

These sequences show what kind of transformations of male infertility take place in the technological practice of IVF. Far from merely applying the technology to a given problem, a practice that is commonly understood to require the technology to adapt to the problem, it is the problem itself which is undergoing various metamorphoses, so that a fit between problem and technology is reached.[161] These metamorphoses establish a trajectory along which the problem in its various guises can be seen to shift: it moves in and out of different bodies, within bodies from one part to another, passing through laboratories and Petri dishes along the way.

First, the problem moves from a male body to the semen. Significantly, the direction of this shift is a move away from the male body to a bodily secretion; the problem does not return at any point in the trajectory to this first body, nor does this shift (usually) involve any medical intervention. The second shift, from semen to conception, represents a shift toward a second body, conception being an event that is supposed to occur in a female body. Subsequently, a very crucial shift follows, the one from a female body to a Petri dish in a laboratory. Defined as a lack of fertilization in vitro, the problem is now located at the point of interaction between oocytes and sperm. Although conceptually this shift represents a move away from the body similar to the first shift involving the male body, there is only the suggestion of symmetry. Materially this shift is conditional upon getting the oocytes into the laboratory, a process that requires a lot of intrusive, hard medical work. It involves a series of various kinds of concentrated moves in and out of the female body, including chemical (hormones, anti-biotics), visual (in the beginning laparascopy, now mainly replaced by ultrasound) as well as a mechanical (needle, speculum, vaginal ultrasound) interventions. The same holds true for the definition of the problem in terms of (poor) IVF outcome, whether this be measured in terms of embryo transfers and implantations, chemical, clinical, and ongoing pregnancies, miscarriages or births.

Each successive shift contributes to the definition of problems that lies at the basis of many of today's efforts to find treatments for male infertility. It is, however, specifically the localization of the problem at the moment of fertilization, i.e. the fusion of oocyte and sperm, that is central to the use of IVF in these cases, as clearly stated in the following quotes:

> In cases of male disorder, the principle of treatment [with IVF] is to *facilitate the contact between oocyte and spermatozoa.*"[162]

> The advantage of IVF, particularly in male disorders, is the *close sperm/oocyte contact* which occurs *under controlled conditions.*[163]

As can be inferred from these quotes, "male disorder" (male infertility) is not a pathology of the male body here, nor even a charachteristic of sperm, but as much a difficulty of oocytes and sperm coming into contact. The problem, as it is defined at these points in the texts, is thus removed from the male body, for neither defective spermatogenesis nor the quality of the sperm is any longer the issue.

Interestingly, the advantage of using IVF in solving the problem, is conditional upon a *definition of the problem* of male infertility *made possible* by the application of IVF in the first place. Consider what counts as the object of treatment in the following quote:

> Human in vitro fertilization (IVF) can be considered a *treatment for aberrant gamete interaction* because sperm and eggs are placed in close proximity under carefully optimized conditions.[164]

'*Gamete interaction*' is a phenomenon that is produced in an IVF laboratory. As an object of investigation as well as of "treatment", it exists only as a *consequence* of IVF treatments.[165] To produce 'gamete interaction' in a laboratory, the availability of gametes, i.e. sperm and oocytes, is obviously required. Although for purposes of research oocytes are sometimes retrieved from patients undergoing sterilization procedures or ovarectomies, the oocytes available in an IVF laboratory are generally from women who have been hormonically stimulated and have undergone ovum-aspirations in the course of fertility treatment. Even basic research of the fertilization process is conditional upon the availability of these "research materials", whose presence in the laboratory already constitutes a phase in some patient's treatment. Thus, to identify the problem of a particular male as a fertilization problem, as opposed to a spermatogenesis or sperm quality problem, his female partner already has to be enrolled in an IVF procedure.

Clearly, however, "optimization of conditions of contact" is not enough to (dis)solve the problem of male infertility. When there is a male problem, it is exactly fertilization in vitro that most often fails to occur. Remarkably, these observations, coming from IVF-specialists themselves, do not lead them to deny the usefulness of IVF in treating male infertility. Rather, IVF has become a presupposed setting for addressing male infertility that generates its own specific definition of the problem:

> The less severe parameters define a male factor in the clinical setting; *the severe ones* are those that have been demonstrated to *cause problems in our embryology laboratory during IVF therapy*, and therefore *they are used to define a male factor in assisted reproduction.*[166]

Clinical criteria for judging the quality of sperm - for diagnosing male infertility - are modified once the problem has been moved to the IVF lab. IVF's definition of male infertility is one that, as the above quote demonstrates, permits exclusion of less severe cases. The percentage of male cases yielding fertilization rates comparable to the average fertilization rates in cases of female infertility makes up the difference between the two definitions. Male fertility problems that do not cause trouble in the IVF lab, in this definition, fall outside the problem range. Consequently, male infertility that still exists in the laboratory setting, is being redefined as well. In the practice of IVF, then, the problem of male infertility is reconstructed. Rather than being a clinical problem in itself, male infertility comes to be equated with a failed IVF attempt. Once the problem is located in the IVF laboratory, the problem is defined as 'aberrant gamete interaction'. Subsequently, it can again be redefined as a 'failed IVF attempt'. The next shift in the localization of the problem moves in one of several possible directions.

A first possible direction is to proceed with the subsequent steps of the IVF-procedure. For example, if the fertilization rate is low because of sperm defects, one can shift one's attention to implantation rates of oocytes that were fertilized. The quoted paper by Diedrich et al.[167], called "Transvaginal tubal embryo transfer: a new treatment of male infertility", provides a nice illustration. The new treatment mentioned in the title consists of a modification of the usual embryo transfer procedure. Instead of transfer to the uterus, a more complicated and invasive procedure is tried, namely a transfer of the embryo into the fallopian tube. So here the problem has shifted once more: no longer located in the male body, in the sperm, or in the IVF lab, it is now situated in the uterus of the woman, where implantation of the embryo is not taking place in a satisfactory way. Since there are fewer successful fertilizations when defective sperm is used, and thus fewer embryos available for transfer, it becomes all the more crucial that those few do implant successfully - hence the idea to change the transfer procedure.

A second possible direction for shifting the problem of male infertility, once it is defined as a failed IVF attempt, consists of zooming in on the interaction between oocyte and sperm cells. In recent years, several techniques have been developed to actively bring about fusion of the gametes, when bringing them together in a Petri dish alone does not result in fertilization. The general terms for these techniques are "micromanipulation" and "assisted fertilization". Among these techniques, three main strategies can be distinguished: injection of one sperm cell directly into the ooplasm (microinjection or intracytoplasmic sperm injection); injection of several sperm cells into the "perivitelline space", that is, the space between the zona pellucida (a proteine rich, transparent "halo" surrounding the oocyte) and the cell membrane (usually referred to as subzonal transfer, subzonal microinjection or subzonal insertion); the creation of an opening in the zona pellucida with chemical or mechanical means, after which several sperm cells are added to the oocyte (usually called zona drilling, partial zona dissection, or zona slitting).

Papers reporting these techniques typically define the problem, and, consequently, characterize the patients as belonging to three groups: "couples who failed fertilization previously, others not acceptable for IVF, and a third group in whom IVF was expected to fail."[168] Patient groups are defined not in terms of bodily dysfunctions, but by their having failed previous IVF attempts, or by the anticipation of IVF failure. Although elsewhere the article makes clear that "couples with severe male factor infertility" are the focus, in the patient definition no reference is made to the male fertility problems themselves. Patients are characterized in terms of IVF success and failure. In addition to identifying the patient in gender neutral terms of "the couple", the definition of the patient's problem is also reconstructed in gender neutral terms.

Underlying the relocation of male infertility to the IVF lab is another important shift. Establishing "ideal", i.e. observable and controllable, conditions of contact between oocyte and sperm cells is not the only way IVF works. IVF also improves the odds of fertilization by a sheer increase in numbers of available oocytes. Whereas normally a woman produces only one oocyte per month, the process of overian stimulation through administration of large doses of hormones makes several oocytes available at one time. The numbers vary enormously, depending on unknown individual factors, but probably also on differences in the stimulation protocols employed.

This shift constitutes a displacement or *condensation of time*: if in one hormonally stimulated cycle plus ovum aspiration you manage to get, for example, 12 oocytes (this number is not uncommon), you have managed to concentrate the chances of fertilization of one year into a single month. Although this is never mentioned as such in the papers reviewed here, it is an important principle underlying IVF in general. In the case of male infertility, however, it signifies a special kind of shift: a reversal. In these cases the reduced numbers of available good quality sperm cells are compensated for by increasing the number of oocytes.

The importance of this principle of changing the odds in the treatment of male infertility becomes evident when you consider the sometimes staggeringly high numbers of oocytes made available in order to establish a fertilization. A few examples from the literature show numbers such as: 1241 oocytes from 166 women (leading to 177 embryo-transfers but only one live birth),[169] 2241 oocytes from 175 women (114 ET's, 24 pregnancies),[170] 590 oocytes from 43 women (34 ET's, 4 pregnancies),[171] and 245 oocytes from 22 women (16 ET's, no pregnancies).[172]

One of the most extreme examples of this type of shift was a study of five women with infertile partners who were hormonally stimulated to produce 110 oocytes,[173] representing an average concentration of the chances of 22 months. The authors of this particular paper stated that, in view of the fact that eventually all of these women delivered a child, though the husbands had "extremely impaired" spermatozoa, their approach must be encouraged, "irrespective of the quality of the semen".[174] Reports like these almost make one feel that, if it were only possible to multiply the numbers of oocytes without limits, male infertility would be a thing of the past.[175] It is perhaps worth mentioning here that one of the most dangerous complications of IVF lies in inducing such "superovulations", for it can result in the occurance of "hyperstimulation syndrome" that in extreme cases can be - and has been - fatal.[176]

I have shown that the notion of the couple as the patient in male infertility treatment is produced and sustained in specific ways. It takes several transformations and shifts in the *localization* of the problem to consider male infertility a diagnosis pertaining to couples and IVF an appropriate treatment strategy of this couple's problem.

We saw that the problem is almost immediately taken out of the male body by characterizing it in terms of properties of sperm. Crucial to the shift to the female body, however, is the localization of the problem at gamete interaction. The translation of the problem of male infertility into a lack of fertilization of oocytes is based on the possibility of medically investigating this process. Thus the very (re-)definition of male infertility *presupposes* the IVF technique and the manipulation of the female body, for it requires moving fertilization and oocytes from the female body to the laboratory. Once the problem has been moved to the laboratory, further transformations can take place, such as shifting the problem along the sequence of subsequent steps of the IVF procedure, for instance, locating it in the embryo transfer and implantation phase. To solve the problem thus conceived, alternatives in transfer procedures as well as additional hormonal treatments are applied. Another direction is exemplified in the fast growing practice of micromanipulation, in particular the variety of these techniques currently becoming publicly known as ICSI (intra-cytoplasmic sperm injection): a shifting of the problem along a path opened up by zooming in on events in the Petri dish. Next to shifting the location of the problem, a secondary mechanism was identified in the changing of the odds of fertilization achieved by reversal of the numerical ratio in available gametes. The scarcity of good sperm is partly compensated by hormonally manipulating ovaries to develop several oocytes in one cycle instead of one. Thus, time is condensed, by concentrating opportunities dispersed over several months into one moment.

3.2 *Working from Bellies to Babies and Back*

In turning to the construction of the fetus as patient in fetal surgery, we find that similar processes of problem transformation are at work. However, the main mechanism involved in transforming the problem of congenital disease differs from the shifting between bodies and locations which, as we have seen, was the case in male infertility. To get a handle on the particular mechanism involved, it is instructive to look first at what happens to the position of 'women', when their role as patient is taken over by the fetus (as was described in the first section of this chapter).

The fact that women usually are not included in the category 'patients' leaves open the question how women do figure in the practice of fetal surgery. It is, after all, hardly conceivable that they are not in some way or other acknowledged to be present. In fact, the feminist observation that a focus on the fetus reduces women to empty background is only part of the story of the construction of the fetus as patient. The presence of women is (sometimes) acknowledged, but neither as patient nor "fetal environment". Instead, they seem to have become *the patient's mother*. The following quotes are characteristic:

> Frequent fetal heart monitoring was performed, and when fetal behavior became normal, the mother was permitted to go home.[177]

> The mother and fetus are sedated to allay maternal anxiety and reduce fetal movement.[178]

The most immediate discursive effect of the term 'mother' is that it distinguishes the woman from the fetus. The addition of the adjective 'maternal' preceding a physiological process or body part is the accepted way in perinatal medical discourse to refer to women's physiology or anatomy. This distinction between 'mothers' and 'maternal factors' on the one hand, and 'fetuses' and 'fetal factors' on the other, is carried through in fetal surgery to the point of becoming absolute. Just as interventions on the mother are thus distinguished from interventions on the fetus, so are effects of the procedures on the mother (if represented at all) distinghuished from effects on the fetus. By implication, interventions and effects on fetuses are *not* interventions on women, and vice versa. Thus, while the mother may be given an anaesthetic, it is not she but the fetus who subsequently undergoes "shunt placement". Conversely, while the fetus may deteriorate and die after or even from the procedure or the preterm labor induced by it, the mother may be said to suffer no complications or morbidity.[179]

Another, more subtle effect of referring to women as 'mothers' is that their position with regard to the fetus, defined as 'other', becomes characterized in terms of *family relationships*. Besides referring to a biological relationship, the term 'mother' is also inextricably bound up with a host of socio-cultural meanings and connotations having to do with norms, values and expectations. The transformation of women patients into the mothers of the patients increases the susceptibility of the practice of fetal surgery to other meanings culturally associated with motherhood, like responsibility, taking care of the needs of others, and self-effacing behavior.

The pervasiveness of the model of family relationships in the context of pregnancy and the medical care surrounding it is remarkable. It ranges from physiological descriptions to remarks about counseling and consent procedures. Since women are not represented as the patient but as mother of the patient, it is *as a parent* that she is counseled and consenting to the surgery. Accordingly, it is actually not she, but *'the family'*, of which she presumably forms only a part, from whom consent is obtained. In practically each instance where choices, decisions, counseling, and/or consent procedures are mentioned, 'the family' or the plural form 'parents' appears:

> *The three patients* treated prenatally by in-utero transplantation of human fetal liver stemcells *were two fetuses* with severe immunideficiency diseases *and one fetus* with beta-0 thallassaemia. *Informed consent from the parents was obtained* prior to initiation of any treatment.[180]

> Over five days, the oligohydramnios continued to worsen and fetal bladder distention increased. *The family was counseled* about the options available including termination, observation, vesicoamniotic catheter shunt, and open fetal surgery. After extensive discussion of the risks of hysterotomy and fetal bladder marsupialization, and the experimental nature of the procedure, *the family chose to undergo open decompression.*[181]

> In four *fetuses*, irreversible renal and pulmonary damage was predicted and *the families were counseled.*[182]

This results in the peculiar situation that in a majority of the texts, we find no 'women' present as embodied individuals undergoing surgery and doing so by their own choice: the patients are 'fetuses', for whom choices and decisions are made by the vague collectivity of 'families'. As in the case of born children, the 'parents' (presumably a woman *and* a man) are construed together as *decision makers by proxy*, on behalf of their still incompetent offspring.

So, women *are* present: as mothers or family member of the patient. This model of a family relationship points to the nature of the mechanism involved in the transformation of congenital diseases (of children) into pregnancy problems (of women). One may call this mechanism *"time displacement"* or displacement of time-frames. Fetal surgery can then be described as the result of a transformation process of congenital medical problems of neonates and children; this transformation process, that in itself is attributable to a diversity of heterogeneous factors, consists of shifting those problems along a time-axis. For example, by referring to the pregnant woman as the "mother" of the patient, her position vis-à-vis the fetus is construed not only symmetrical to that of 'the father', but also analogous to the *post*natal situation. After all, elsewhere we reserve the term 'mother' for women who have given birth, and think of pregnant women as persons about *to become* mother. Similarly, the repetitive mentioning of families deciding on treatment and giving consent, suggests that a model deriving from the *post*natal parent-child relationship is at work in the distribution of responsibilities and authority.

While the moment of birth is the most graphic image of individuation we have,[183] medical conceptualization of the period preceding that moment consists of importing language and meanings from a phase that, relative to the pregnancy phase, can only count as the future. But this conceptualization, however consistently adopted, and though functional and innocent in some contexts, cannot but lead to frictions - and sometimes plain absurdities - in others. Thus, once in a while fetuses are described as having congenital diseases, or even "in*born* errors", whereas my dictionary says 'congenital' to mean "present at birth", and fetuses, whatever else they may be, are certainly not born yet. Conversely, the mixing up of time-frames surfaces when fetuses apparently at times can be thought of as "postnatal" (as in "fetuses receiving postnatal treatment"). Confusion about the relation of pregnancy to (future) individuality is further evidenced in funny mistakes like the following:

> *The mother and the fetus* had an otherwise uneventful recovery; *they left the hospital* on the 14th day and returned to a small town in New Mexico to be followed by her local obstetrician.[184]

But the significance of "mistakes" like this is that they indicate how meanings derived from an anticipated future may shape the discursive development of fetal surgery. In order to further explore how this mechanism of time-displacement works with regard to the transformation of problems, we will now take a closer look at the role of the time dimension in the definition of the problems dealt with by prenatal intervention.

First, there is a sense in which fetal problems and their severity are characterized and named in terms of the projected and anticipated problems of the children they may become. This, of course, has a certain obvious logic to it, in the sense that many prenatally detectable anomalies are good predictors of postnatal viability and potential problems and diseases. However, the essentially prognostic value of many prenatal diagnoses can subtly slip into the attribution of conditions, diseases, and specific forms of suffering to the fetus:

The case reports of the first two *patients treated* by in-utero FLT [fetal liver transplantation] are summarized below. Both of them *were fetuses* with severe immunodeficiency diseases diagnosed in mid-gestation. The first *patient suffered* from bare lymphocyte syndrome,   ...   Infections,   especially   with   opportunistic microorganisms, are responsible for death of *these infants*, unless they grow up isolated in a fully sterile atmosphere while they are recon-stituted with stemm cell transplants.[185]

The natural history of congenital obstructive uropathy mandates the need for improved therapy earlier in gestation, in order to salvage *fetuses who would otherwise die of renal failure and pulmonary hypoplasia.*[186]

One may ask in what sense a fetus can be said to suffer from immunodeficiency disease, or to die from pulmonary hypoplasia (underdevelopment of the lungs). In instances like this, the condition of a fetus seems to be characterized and judged as if it were already born. This slipping between meanings and events connected to different time frames is especially clear in the first quote, where the fetuses are literally referred to as "these infants", who can also be seen to be "growing up". Similarly, the second quote has fetuses dying of underdevelopment of the lungs, whereas fetuses, unlike born children, do not depend on their lungs for oxygen. A detected determinant of future illness, in a sense, acquires the weight of the illness itself. A possibility of future suffering is retrojectively turned into fetal suffering in the present.

More frequent, as well as consequential, are rationales and justifications for surgery during pregnancy, in which fetal surgery is construed as *prevention*, as a considerable number of practitioners and researchers do. As one surgeon phrased it: "Advances in diagnostic and surgical techniques have provided a new basis for prevention of certain congenital defects by intrauterine therapy."[187]

The emphasis on the 'preventive' character of what in themselves are invasive procedures underscores the centrality of the time dimension and the mechanism of time displacement. The very meaning of 'prevention' is to do something *now*, in order to avoid something worse to happen *later*. The term refers to the future: its making sense rests on the invocation of an image of a likely but undesirable future, that in the same movement is rendered avoidable. Indeed, the use of the concept of 'prevention' has a strong rhetorical effect, with distinctly positive connotations.

The frequency of this use of the rhetoric of prevention in this context is another indication of the role played by a specific conception of time. Underlying these examples is the construction of a *continuous* time-axis, along which medical problems of born persons are retrojectively shifted in time. As could be seen above, fetuses can be equated with born infants, as if there were no significant moment of transition in which fetuses first *become* infants.[188] A similar kind of deconstruction of the relevance of birth is exemplified in conceiving of pregnant women as mothers. Discontinuity between the prenatal and the postnatal stages is replaced by a discursive construction of continuity. The centrality of this continuous time-dimension is further underscored by the ubiquity of words like 'before', 'too late' and 'earlier' in rationales for intervention given, rendering fetal surgery the "earliest possible pediatric surgery." More specifically: they show how fetal surgery is premised on a type of reasoning that represents pregnancy exclusively as the early part of the lifetime of the future child, at the expense of its visibility as part of a woman's life.

This perspective becomes somewhat understandable, if one considers the medical specialties from which fetal surgery has evolved. Allthough fetal surgery involves a team of specialists from a variety of medical disciplines,[189] it is above all pediatric surgeons who are pushing the development of this new field. Next to pediatric surgeons, plastic and reconstructive surgeons are the ones researching the possibilities of in utero repair of cleft palates; similarly, the 'prevention' of hand deformities is foreseen by an orthopaedic surgeon. Unlike gynecologists and obstetricians, these specialists clearly are not primarily involved with women as their patients. Their work is in a field in which they are confronted with the tragedy of neonates and children entering the world in a sick or deformed state. It is conceivable that frustration about their often doomed efforts is experienced as 'being too late', even when they are present at the very birth of their patient. From their perspective, they are always late, in the sense that they see the problems they are confronted with as the end result of a developmental process gone wrong. To them, in utero surgery represents a possibility to reach their patients "earlier" and "prevent" the damage. Understandable as this may be, it is also clear how women have never been central in their efforts, but were encountered as their patients' mother. Although there is an awareness that, when moving from pediatric and neonatal surgery into fetal surgery, the "mothers" encountered stand in a somewhat more complicated relationship to the fetal patient than to the born patient, there is a tendency to regard the problems associated with this complication as secondary to the main problem. It practically even belongs to other disciplines, in particular to those covering Cesarean section, ethics, and law.

It would be unjustified to suggest that the female bodies are lost from consideration in this practice; indeed, many authors testify to their acute awareness of the difficulty of judging when it is, medically speaking, 'required' to submit women to procedures explicitly conceptualized as benefiting 'someone else', while carrying significant risk to them. What is suggested here, however, is that, despite such awareness and a probably sincerely felt concern, there are mechanisms involved in the particular way the fetus is made into a patient that systematically add to the effect of making ever more interventions look justified and magnifying perceived advantages.

Adding to these effects is the specific quality of the time dimension underlying the shifting of medical problems and their treatment back into the prenatal stage. Besides a construction of prenatal and postnatal time as as continuous, it is also *non-linear* time. We may call this mechanism the *condensation of time.* The strong emphasis on developmental processes involved in pregnancy, produces a perspective in which the time before birth acquires significance and importance seemingly never equalled again in the post-natal period. The embryological, developmental perspective carries strong overtones of teleological thinking, in which the growing of the fetus is perceived as a process of unfolding of what is in principle (in potential) already there. Sarah Franklin points this out when suggesting that:

> The emphasis upon what the fetus is going to become, upon its genetically determined development, inevitably leads to a focus upon its developmental potential as a person, as an individual human being with an entire life course mapped out for it from the moment of conception. ... There is thus a sense in which the conceptus is provided with an entire life cycle through the construction of its developmental potential, which is simultaneously naturalized and authorized through its representation as biological fact.[190]

Consequently, an intervention during pregnancy is never just the mending of a dysfunction or pathology in the present, but is carrying the weight of mending an entire future lifetime. In this sense, pregnancy is not even just the first nine months of a persons life, but *contains* the (say) ninety next years in a condensed form:

> The objective of fetal surgery is to reverse or arrest a destructive *process before* irreversible damage occurs.[191]

> Even though this technique is still in its infancy and its efficacy has not yet been clearly established, the assumption is that such treatment under appropriate circumstances *should give the affected fetus a greater chance for a reasonable life.*[192]

By constructing the fetus as the patient, pregnancy is not only increasingly and exclusively seen as the early period of a future person's lifetime, at the expense of its visibilty as part of a woman's life, as stated above, but there is also a mechanism at work that makes the nine months seem incomparably more important for the fetus than for the woman. For her, pregnancy does not come to be perceived as constituting 'condensed' time but just 'normal' time. Even when medical interventions would be conceptualized as involving women as well as fetuses, their stakes would be defined in uneven terms. Everything less than life-threatening or permanently incapacitating for the woman may come to count - when set against the gains perceived in terms of entire lifetimes - as temporary discomfort or reasonable risk. How will three weeks in hospital be weighed against ninety years saved?

Another aspect of this teleological perspective, is that it carries the possibility of *infinite regress*. The potential-actualization distinction is not in any way 'naturally' bound up with the prenatal and postnatal phases. In a developmental process, any phase can be construed as an actualization or unfolding relative to the phase preceding it. Conversely, any, or even the same phase can be perceived as already containing all the seeds or potential for the one superseding it. In this way, such developmental, teleological vocabulary can be applied arbitrarily to the point of becoming mere rhetoric. The only constant, in the end, is the discursive effect of assigning more importance to one phase relative to the next.

This possibility of infinite regress is nicely illustrated in the following quote:

> However, the potential advantages of correcting a deformity before birth are impressive, for a mechanical obstruction can be bypassed *before* it causes severe secondary changes. ... We might reflect upon the benefits of such *preventive surgery for congenital deformities of the hand*. For example, consider the advantage of releasing the annular bands *before* the tissues distal to the constriction become swollen or gangrenous. The severe hand deformities that often occur in the wake of these constrictions could be *prevented*. But there are other advantages to fetal surgery that have even greater importance. They are as follows: (1) The *potential* for healing and regeneration are many times *greater in the fetus than in the fully formed infant*. Surgical wounds would heal rapidly - some without suture. ... Thus, in many ways, fetal surgery may become safer than operating on the infant.[193]

There is hardly a sense in which an infant can be said to be "fully formed". One can safely presume that most adults, since infancy, have managed to develop beyond a form of life restricted to lying around in cribs sleeping, crying, sucking and digesting milk. Even biologically speaking, it is unclear what could be meant with "fully formed infants", since in biological terms as well, infants continue to develop. The main effect of talking of "fully formed infants", by contrasting them to the fetus' potential, therefore, is producing a rather absolute sense of being *too late* at birth, and designating the fetal stage as the stage where it makes most sense to intervene.

A similar kind of reasoning is exemplified in most texts, usually in a more subtle way though. To give one more example, in which an argument is put forward against *post*natal treatment of a particular affliction:

> It is clear that the lung made hypoplastic by CDH [congenital diaphragmatic hernia] can grow and *develop after* it is decompressed *at birth*, but the *potential* for further growth is limited by the relatively *late timing* of decompression."[194]

While an adult can be seen as the actualization of the potential of an infant, it is easy to see how this aristotelian dichotomy can be arbitrarily shifted to produce the "fully formed" infant as the actualization of the potential of the fetus. The possibility of infinite regress is nicely illustrated by one of the same texts quoted above, whose author enthusiastically continues to phantasize about a future in which total bio-engineering is practiced as the ultimate prevention. Notice how, in passing, he invents the "full grown fetus", in iuxtaposition to the potentiality of embryos and genes, to the rhetorical effect described above.

> If the genetic code can be read, it will not be long until it is decoded. Soon thereafter the molecular engineers will be able to decipher the genetic code from the cells of the developing embryo and predict the features of a *full grown fetus*. By recombinant DNA techniques already in use, they will be able to remove offensive base pairs destined to produce deformity, and then splice into the DNA molecule base pairs that will develop normal tissues. In other words, scientists will be able to *correct* genetically predetermined defects *before they occur*. Before long, genetic manipulation, in vitro fertilization, and sperm selection will *prevent* many defects *before deformities develop*.[195]

In this section I explored some of the mechanisms involved in the transformation of children's congenital problems into pregnancy problems. Whereas the previous section on the transformation of male infertility identified a shifting of the *location* of the problem as the central transformative mechanism, the case of congenital disease foregrounded a comparable but different mechanism. We saw that a *temporal* dimension, instead of a spatial one, plays a determining role here. Pregnant women are anachronistically transformed into 'mothers', members of the family of the fetal patient, whom they represent as a parent; the hitherto mainly prognostic value of prenatal diagnosis is turned into indication for intervention in the present, which, in turn, is conceived of as prevention and advancement of treatment to an earlier point in time. Thus congenital problems can be seen to transform as the result of shifts along a time axis construed as continuous with the postnatal period.

In addition to its construction as continuous, the dominant developmental perspective on the fetus also produces a rendering of the prenatal period as non-linear, or condensed time. By conceiving of each phase in fetal development as containing all that is necessary for its developing into the next phase, the prenatal phase is weighted with an importance far greater than the postnatal period, or the same period as a phase in an adult woman's life. This construction of a time axis allowing for the transformations surrounding congenital disorder is not merely a matter of changing conceptualizations, but it is actively and materially produced in sequences of medical - often "diagnostic" - interventions in female bodies.

4. COUPLES AND FETUSES AS HYBRIDS

In the previous sections, I explored how the technological practices of in vitro fertilization and fetal surgery transform the problems they supposedly treat. This is a departure from the commonly held belief that using new technologies for old problems, or existing technologies for new problems, merely involves adaptation of the technologies to adjust to pregiven and pre-existing problems. In the process of transforming the problem, I argue here, they simultaneously create the patients that are identified as the ones being helped with the technologies. By this I do not mean that the problems treated are iatrogenic in nature, but something very different. My analyses suggest that the particular medical interventions and actions involved in the technologies of in vitro fertilization and fetal surgery produce the 'patients' named as the bearers of the problems treated, specifically, couples and fetuses. Their emergence as patients cannot be seen apart from the specific pathologies they suffer, because it is these pathologies that define their identity. In being construed as *patients*, the couple in infertility treatment exists *as* infertile couple; the fetus in fetal surgery exists *as* diseased or abnormal fetus. Their pathologies, in turn, cannot be understood separately from the technologies which transform and define them, render them visible and treatable.

In seeking to comprehend how women's bodies have become the designated location to solve problems that used to be children's and men's, one should first understand how the answer that such problems are, actually, the problem of couples and fetuses is an obscuring one. In fact, what these analyses point to is exactly the reverse. Through a change of perspective, that Katherine Hayles calls 'reflexivity', it became possible to identify a clear instance of the way "scientific experiments produce the nature whose existence they predicate as the conditions of their possibility."[196] We do not operate on women for these problems because it was discovered that it is really couples and fetuses that need treatment, a discovery finally showing the nature of the problem to necessitate interventions performed on women; rather, it is the other way around. It is *because* we operate and experiment on women (and have a long history of doing so) that we now have fetuses and couples as patients; they originate in a specific tradition of interventions involving women's bodies.[197] In order to see fetuses and couples as individual entities capable of posessing properties like being infertile or diseased, we have to intervene in women's bodies.

Long sequences of interventions are in fact necessary before one can produce the phenomenon of "gamete interaction", as something observable and amenable to treatment in a laboratory. A woman has to go through the so-called "stimulation protocol", an intervention that first puts her menstrual cycle at a standstill to enable an artificial, controlled take-over that makes her ovaries do something they would not do otherwise produce multiple ova at the same time. This event is timed and triggered to fit a schedule convenient for the planning of the next intervention, the "ovum pick up", which is a highly technologically mediated procedure as well. These artificially produced ova then require very tenuous and painstakingly created environments and equipment to be able to survive, be seen, and manipulated. The semen to be added has to go through some transformative procedures as well; it is "washed" and centrifuged, before being joined to the ovum in the dish. Only now, with the additional help of a microscope, do we have the object of treatment called 'gamete interaction', a singular phenomenon, where the individuality of the two bodies of the couple concerned does not play any material role. After all this creative work that, as far it is directed at living, material bodies, concerns almost exclusively the female one, and before the next series of similarly directed interventions, this is the moment when a couple emerges as one singular medical patient.

Similarly, a long path of medical interventions on women's bodies has to be followed before the patient in fetal surgery acquires the material reality it has on the operating table. Starting with the by now familiar visualization of her insides by standard ultrasound, the woman has to present her body patiently for various types of inspection which - when something is suspected to be wrong - are likely to grow in invasiveness, painfulness, duration, and personal risk. In the process of advancing along this path, all the while the fetus becomes more clearly circumscribed and pictured, acquires more defining characteristics (in the form of specified diagnoses), and so, gradually, turns more and more into a real "patient". To establish whether this fetus will become a surgery patient, it has to be ascertained, as far as possible, whether this fetus only has the problem the intended procedure might amend, and not serious additional ones that would render the attempt futile, and whether its condition really warrants intervention before rather than after birth. Following Stefan Hirschauer's argument,[198] the surgical procedure itself, then, can be considered the finalization of the material creation of the fetus as patient. Ultimately, it is in the anesthetization of the woman, the connection of her vital functions to the monitoring equipment, and the cutting through the various layers of her body, that the fetus is materially produced as a surgical object of treatment.

In view of all this, it seems hardly adequate to consider couples and fetuses natural entities, in the sense of being simply or naturally there, waiting, just until this historical moment, to be discovered, observed and ultimately, treated. Considering all the highly skilled work that has to be done, all the sophisticated technologies that have to be in place, and the careful orchestration of the many elements constituting the environment in which they can come into existence, they are at least as artificial as they are natural.

Obviously, this argument can be extended - as it has been - to any object of scientific investigation. It holds true as much for any elementary particle discovered at CERN, HIV-virus,[199] or Neanderthal skeleton, for that matter. Without the most sophisticated dating technology even a valuable prehistoric find remains a few scattered bones. So, being what one might call a hybrid of nature and technology,[200] is not a unique characteristic of the particular objects, fetuses and couples, reproductive technologies have produced. The latter differ, however, from those other hybrids in crucial ways.

What makes couples and fetuses so special, is that they are not just 'objects' but - indeed - 'patients'. That is to say, they occupy positions in medical discourse akin to human beings in significant respects. While any doctor would deny they are full-fledged human persons (at least in the case of couples; with respect to fetuses more extreme views exist), they acquire so many characteristics comparable to human persons or subjects in these texts, that it is sometimes difficult to assess whether or not differences are still recognized. Like human persons, they suffer from health problems and have specific needs, they are diagnosed and have prospects, they come into clinics and receive treatments. What sets them apart from mere 'objects' most definitively is the consent required before anything may be done to them.

But both couples and fetuses transgress conventional notions of what can count as an individual patient. To the extent that they count as *one*, *individual patient*, they depart from (at least) one fundamental attribute 'other' individual patients possess, namely to be *embodied* in a way that gives coherence to the concept of counting as an individual patient.[201]

As we have seen, their very existence as patients is intrinsically connected to particular redefinitions of medical problems. The processes involved in redefining these medical problems simultaneously generate the 'patients' to whom these problems are said to belong. However, we have also seen that the processes of transforming these problems are conditional upon medical interventions in the bodies of individuals *not coextensive* with the ones designated as the patient. To transform the problem of male infertility into the problem of a couple, it takes medical work on the female body (mostly). Similarly, the female body is the object of the interventions presupposed in transforming congenital disease into a prenatal problem of a fetus. One cannot consider fetuses or couples as discrete, bounded patients, *without* implicating the boundaries constituting an individual female body as well. Thus, fetuses and couples form an anomaly with respect to a medical notion that, up until now, was undisputed and considered self-explanatory, and therefore hardly required being made explicit, namely that a medical patient is one individuated, embodied person. While 'individuality' knows only the states of being one or two, as mutually exclusive categories, fetuses and couples are neither one nor two, or perhaps both. They are, in this second sense as well, hybrids.

CHAPTER 3

TREATMENTS FOR MEN AND CHILDREN

1. INTRODUCTION

Crucial to the analysis of reproductive technologies is the recognition that these practices do not only involve the blending of nature and technology, but also that they constitute discourses on the production of *individuality*.[202] This is one of the reasons why "reproductive politics" has been and still is central to women's emancipatory struggles. Pregnancy is about the making of new individuals, about processes of individuation, about two bodies becoming one, one body becoming two.[203] Individuality, however, whether understood psychologically, morally, legally, or even biologically, is not a pregiven ontological category, but always a contingent achievement.[204] At the same time, it is fundamental to most of our notions that invoke normative issues in medicine, such as patient autonomy and bodily integrity; it underlies patient rights and informed consent procedures in medicine. With respect to bodily self determination, for example, it is obviously required that it be clear what counts as self, and what as other, where the boundaries of the individual body are drawn. In contemporary reproductive technologies, however, it is precisely these boundaries that are at stake and being redefined.

By focusing on two fields within reproductive medicine and technology, infertility (in vitro fertilization) and congenital disease (fetal surgery), the emergence of two new, extraordinary types of patients was analyzed in the previous chapter. Extraordinary, because they depart from conventional notions of what can count as an individual patient. 'The couple' in infertility treatment (male infertility in particular) and 'the fetus' in fetal surgery have come to be considered independently identifiable and treatable, single patients. Significantly, they have emerged as such in contexts where women now are being medically treated for problems that used to belong to others, that is, for problems that used to be their children's and male partners'. Moreover, this development challenges any self-explanatory use of the notion of a patient's bodily self-determination in medicine, because it occurs in technologically induced clinical contexts where it is no longer clear *which* selves or *whose* bodies precisely are involved and to what extent, nor is it even always possible to say *how many* selves and bodies exactly are involved.

Both couples and fetuses are construed as patients in the very process of defining and transforming the problem from which they are said to suffer, and of which technology is supposed to relieve them. They thus constitute two highly consequential instances of the way reflexivity helps us see how "an attribute previously considered to have emerged from a set of preexisting conditions is in fact used to generate the conditions,"[205] In the process of shifting the problem of male infertility spatially in and out of bodies, body parts, laboratories, and Petri dishes, 'the couple' appears as the new bearer of the problem thus conceived. Similarly, congenital disease can be seen to transform according to changing temporal designations of the occurance of the problem. From being a problem of born children, congenital disease has shifted into the prenatal period, where a 'fetus' is now considered a patient indicated for therapy. These transformations are achieved through elaborate medical procedures, consisting mainly of interventions in female bodies. Again, it is the scientific experiments themselves, that "produce the nature whose existence they predicate as their condition of possibility."[206]

In this chapter, the analysis of 'the fetus' and 'the couple' as 'hybrids' produced in technological practices, is taken one step further. I describe how the notion that fetuses and couples *are* patients is sustained and made durable through scientific accounts that present the construction and transformation processes described in the previous chapter in reversed order. Instead of considering medical interventions as the material preconditions from which new problem definitions and new patients emerge, these texts retrospectively present the medically defined problems of fetuses and couples, as well as these patients themselves, as pregiven phenomena. The problems and the patients are thus seen as the unproblematic starting points from which interventions follow, rather than the other way around. In order to achieve this reversal, the traces of the interventional work necessary to establish male infertility as a couples' problem, and congenital disease as a fetus' problem, are erased from the accounts. I will show how the ambiguous status of these new patients with respect to both naturalness and individuality is resolved by specific discursive patterns.

The first section describes a type of pattern that, following Star (1992), is referred to as 'deletion'. It concerns a class of discursive mechanisms that accomplish the *erasure of interventions* on female bodies, and, by implication, their constitutive role in establishing new patients and new problem definitions. The second section deals with another pattern that further enhances the idea that these technological practices are not about 'women' or 'female bodies'. This 'purification' pattern, as Latour (1993) has called it, reduces the duplicity of fetal patients and infertile couples. Instead of being about women and men as couples, and about women and future children as in fetal surgery, this pattern shifts the tenuous balance by suggesting that these practices are actually still, above all, about *men* and *children*. This makes recognition of the shift of medical problems between 'individuals' significantly more difficult (for, as will be taken up in chapter four, it diminishes the possibilities of seeing individual female bodies being involved in these technological practices at all). A further illustration of the working of the mechanisms involved in these two patterns is given in the third section on scientific evaluations of the technologies. Finally, the chapter concludes with some remarks on the possible effects of these patterns on the changing position of the female body in medico-technological reconfigurations of reproduction and the potential for contesting these changes.

2. THE DELETION PATTERN

An often observed characteristic of scientific discourse is that it produces a sense of neutrality, objectivity, and naturalness for the facts and objects it describes through what can be called "cleaned up accounts".[207] In the first chapter several arguments and examples were given in order to highlight this function of style in scientific writing. Here, I want to focus in more detail on one form of such stylization that is specifically relevant to the politics of science and technology. The standardized formats and economic use of language so typical of technical, scientific writing accomplish effects that go beyond verbal parsimony. Purporting to give only scientifically relevant results, a highly stylized rendering of the work is achieved by leaving out most of the practicalities, day to day contingencies, and details that constitute the larger part of doing scientific research. Thus, in its representational practices,[208] science - and experimental medicine is no exception here - produces its results, its discovered facts and objects, through inevitable selection of relevant details from a sea of irrelevant ones, by making disctinctions between trivial practicalities of the experimental set up and significant theoretical or methodological advancements and results, and so on. Of course, any writing or reporting must necessarily be selective and make distinctions between what is relevant and what is not. Therefore, this selectiveness as such is productive: without it the very possibility of giving informative accounts would be lost.

However, this inevitable selectivity also means that there is no such thing as neutrality in writing. The criteria according to which one distinguishes between the relevant and the irrelevant, between information and noise, constitute a specific perspective, one that inevitably excludes others. From such other standpoints, particular selections of what is to be deleted and rendered invisible may come to look far less innocent than mere practical detail. Star (1992), for instance, convincingly argues that it is also the entire social and political constellation (divisions of labor, distributions of resources, and so on) from which the scientific pursuit of natural facts proceeds, which is erased in this selection. Thus the purity of scientific result can be seen to stem from an unavoidably partial selection of what will be related, from an inevitable complex configuration that is anything but pure or power-neutral.

In this section I focus on some of the 'deletions' involved in constructing couples and fetuses as patients in high-tech experimental reproductive medicine. I focus in particular on practical and material 'details' concerning the way in which female bodies are implicated in these practices.

A first point in this respect is the close interrelationship between the all-pervasiveness of the deletion pattern and the fact that 'couples' and 'fetuses' are considered the patient (in IVF and fetal surgery respectively). This translates into an immediate pattern in which interventions are not described as interventions on female bodies, but instead as interventions on said couples and fetuses. Referring to the fetus as the one undergoing procedures in fetal surgery is by now the standard way of describing things, despite the fact that women are physically involved here. Similarly, in IVF it is couples who are said to undergo the various invasive procedures involved in the technique, a way of speaking that replaces descriptions of women as the patient undergoing procedures. Especially in procedures that involve female bodies only, however one may define 'others involved', it is at first sight puzzling why the interventions should be referred to as follows:

> Intratubal embryo transfer was carried out in 95 *couples* in whom
> male disorder was the main reason for infertility. All patients had had
> at least three intrauterine inseminations before they entered the IVF
> programme.[...] Four main schemes of ovarian stimulation were used
> in these *couples*.[209]

Obviously, both stimulating ovaries and transferring embryos into fallopian tubes is carried out on female bodies, something which seems unnecessarily obfuscated by ascribing it to 'couples'. But this pattern of redescribing interventions on women as interventions on 'other' patients, does not stand alone. It is accompanied by other, less obvious versions of the same mechanism that, together, produce a consistent, cumulative effect. This effect, that probably was never consciously intended by anyone, may nevertheless be quite consequential.

The several varieties in which the pattern occurs have something in common. They achieve the deletion of the involvement of female bodies through *reconceptualizing intervention*: the actual interventions are redescribed and transformed into something else. This concerns less the fact that it is female bodies that are involved, as in the first example of the pattern given above, but rather the interventional character of the procedures themselves so that they are no longer visible as (part of) the therapeutic efforts proper.

A phenomenon constituting a major reconceptualization of therapy is the conceptualization of (major) surgery on pregnant women for congenital disease of future children as *"prevention"*,[210] as is unambiguously stated in the following quote:

> Advances in diagnostic and surgical techniques have provided a new basis for prevention of certain congenital defects by intrauterine therapy.[211]

Since the term 'prevention' signifies taking early action in order to avoid the occurance of more serious problems that need more drastic interventions later, its appropriateness in this context is rather questionable.[212] However, the point here is not to evaluate the appropriateness of conceptualizing surgery as prevention, by weighing the preventive measures against the harm prevented, but to consider the effect of a conceptualization as prevention in itself. This effect can be described as pulling attention away from the intervention as intervention and redirecting it elsewhere. 'Prevention' invokes an image of something prevented which tends to overshadow and downplay the means of prevention itself; as Ulrich Beck writes: "The center of risk consciousness lies not in the present but in the future"[213] Thus, the projected dangers of the future, the images of potential harm, are highlighted in a way that makes the interventions in the present become relatively shaded: preventive fetal surgery is primarily about some future damage avoided. The preventive measures in the present have, as a consequence, become secondary.

In this way the pervasiveness of the vocabulary of prevention in the discourse on fetal surgery suggests serious questions about its limits as an endeavor with a distinctly positive ring to it. Prevention relies on an estimation of risks, it involves the prediction of a future course of events that may then be altered and improved. In the case of pregnancy, establishing the required prognosis or diagnosis is a meticulous process of reducing the uncertainties inherent to every prediction by a growing series of increasingly invasive and risky procedures. Closely connected to the deleting effects of

the reconceptualization of intervention as prevention, therefore, is its
reconceptualization as diagnosis.

The potential of the concept of 'diagnosis' to delete the visibility of
intervention rests on its relationship with the concept of 'therapy'. Together,
these two concepts form a duality that suggests a clean distinction between
the two. Moreover, this distinction confers meaning upon 'diagnosis' as
involving merely the establishment of facts, on which subsequently action is
taken by 'therapeutical' intervention. Thus, the concept of 'diagnosis' invokes
images of mere observation, whereas the image of actual intervention is
reserved for 'therapy'. This picture, of course, does not correspond to what
actually goes on in most of today's medicine generally, where the
development and application of ever more sophisticated diagnostic
technologies, that more often than not are quite invasive, constitute a large
part of the ongoing medical work. Rather than being exceptions, infertility
treatment and prenatal medicine epitomize this character of contemporary
high-tech medicine. In cases of both prenatal anomaly and male infertility,
establishing a diagnosis is a process that - in order to acquire the required
specificity and reliability - involves a growing number of interventions in
female bodies. The following examples, taken from the context of fetal
medicine, illustrate the complexity of the process involved. They also show
the relationship between improvement of information and increase of
intervention and concomitant risk:

> There were no other abnormalities, and fetal bloodsampling revealed a
> 46, XY karyotype. Biochemical test on urine obtained by
> ultrasound-guided needling of bladder and the renal pelvis of both
> kidneys suggested adequate left kidney function. One week after,
> repeat sampling confirmed these findings but showed possible
> deterioration of the left kidney.

> We suggest sampling urine from each kidney rather than from the
> bladder alone in selecting patients for vesicoamniotic shunting.
> Complete information is also obtained by serial sampling.[214]

> Initially, all fetuses underwent a level 2 obstetrical U/S [ultrasound] as
> previously described. Moderate-to-severe oligohydramnios on the
> initial U/S evaluation or progressively increasing oligohydramnios on
> the serial U/S prompted further evaluation of fetal renal function. The
> fetal urinary tract was exteriorized by placing a balloon tipped catheter
> ... percutaneously into the dilated fetal bladder using U/S guidance. ...
> Each mother received intravenous (IV) diazepam and local anesthesia
> prior to catheter placement. ... after maternal hydration ... and an
> intravenous bolus of iothalamate ..., hourly fetal urine output and
> iothalamate excretion were measured. ... Uterine activity was

monitored with an external tocodynamometer, and ritodrine hydrochloride was given if significant uterine contractions occurred. The balloon tipped catheter was usually removed after six hours (range 3 to 16 hours).[215]

These quotes illustrate how the need for adequate assessment of fetal condition leads to quite invasive forms of diagnostic intervention. Besides sampling of fetal blood, these quotes mention how fetal urine is obtained not only from the fetal bladder, but from both fetal kidneys as well. The procedures are repeated or performed in a way that takes hours of catheterization, during which the woman receives several kinds of medication and the occurance of contractions has to be carefully watched and pharmaceutically controlled. One of the most important reasons for caution, according to medical scientists, is exactly this uncertainty of the diagnostic process and the consequent "inadequate patient selection". These are often considered the main factors limiting the success of the experimental procedures. Not surprisingly, medical scientists repeatedly warn against overenthusiasm concerning fetal treatment. Reassuring as such caution and such calls for self-restriction may sound, they do not necessarily mean that there is actually less interference in pregnancies. Seen from the perspective of the patient, however, the distinction between diagnosis and treatment becomes of analytic instead of practical significance. The difference concerns a difference in goal (relieving the problem or gaining information) rather than a difference in measure of intervention. Still, the reconceptualization of interventions as diagnosis can invoke this obsolete image by clearly implying an opposition between the two. In fact, as the next example shows - as do the ones quoted above - such caution can immediately be translated into a need for development and use of more and more diagnostic procedures of various kinds.

> The outcome could improve with further refinements in selection and treatment. Particularly important is the ability to recognize which fetuses cannot benefit from intervention. Chromosomal abnormalities that would have *prevented intervention* in six cases can now be detected without weeks of delay by rapid karyotyping with use of fetal blood, and irreversible renal damage can now be judged by recently developed tests of fetal renal function. ... With the *evolution of diagnostic techniques*, most of these anomalies can now be recognized and *intervention be avoided*.[216]

As is suggested here, "intervention can be avoided" by procedures that constitute, of course, invasive interventions in themselves. To present both fetal blood sampling and fetal renal function testing in opposition to 'intervention' downplays their necessarily invasive, risky character. In the end, it can even become difficult to distinguish 'diagnostic' from 'therapeutical' interventions. What this analysis of the reconceptualization of intervention as diagnosis makes clear as well is how the *construction* of fetal therapy as prevention involves medical intervention in the female body, while *at the same time* it is put forward as the *rationale* for such intervention.

The same redescription of intervention as diagnosis occurs in the context of IVF, where the equation of male infertility with IVF failure corresponds to a changing conceptualization of IVF treatment as well. Whereas IVF was originally considered a treatment for female infertility, seeing it as a treatment for male infertility requires its conceptualization as a treatment for "aberrant gamete interaction." Since as such it is successful in only a small number of cases, it changes quite often from being a treatment into a diagnostic procedure:

> Thus, although the technique of IVF-ET allows an accurate assessment of the fertilizing ability of spermatozoa, which is its major *diagnostic benefit*, the fertilization process itself seems to be the limiting factor governing its clinical benefit in terms of establishing pregnancies. ... *we offer at least one 'diagnostic' IVF-ET to every couple with long standing male infertility.*[217]

> Assisted reproduction [IVF] is *the ultimate diagnostic test* and therapeutic effort for male infertility due to severe sperm disturbances. When failure of fertilization is demonstrated, only assisted fertilization [ICSI etc.] procedures can overcome the problem.[218]

In these instances, it can be seen how the difference between diagnosis and therapy is actually established *post hoc*: one and the same intervention becomes retrojectively relabeled 'diagnostic', depending on whether it failed or succeeded as 'therapy'.

While in practice such diagnostic procedures are likely to be as interventionist as therapeutical procedures, the label 'diagnosis' confers the image that nothing really is done yet, not, at least, in the sense of actively addressing the problem. All that is done is mere recognition and demonstration of the problem. Actual therapy, here "assisted fertilization" (i.e. micromanipulation of gametes), is located elsewhere. By thus (re-)labeling many of the interventions in female bodies required to 'discover' a couples' or a fetus' problem, these interventions have become less recognizable as such.

The next two instances of patterns of deletion are reconceptualizations specific to the IVF context. In examining the parts of the IVF treatment that concern getting the oocytes into the laboratory, that is, the hormonal stimulation of ovulation in the woman and the retrieval of the oocytes from her body, we find another example of reconceptualizing the bodily interventions in the IVF procedure. In contrast with how these stages of the treatment are likely to be experienced by the woman involved, these stages appear to be reduced to the status of *preparation*, necessary to start with the 'real' treatment: the laboratory work. In other words, those aspects of treatment that will be experienced by embodied subjects as being performed on their bodies are excluded from what counts as the treatment in these texts. This exclusion is reflected in the attention devoted to these phases, commonly referred to as "stimulation protocols" and "ovum pick-up". Usually, these procedures are briefly mentioned in a standard, summarizing formulation in the "materials and methods"-section, such as "couples were stimulated with one of the following protocols", reflecting the extent to which these medications and procedures are considered standard and routine. Consider, for instance the minimal reference found in Cohen et al. (1991): "follicular stimulation has been described elsewhere", which literally reduces the manipulation of the female bodies involved to a footnote, while leaving out the ovum-aspiration procedure entirely.[219]

Something similar occurs at the other end of the laboratory phase, where placing embryo's in a woman's uterus has come to denote a *measure of success* in the treatment of male infertility. In tables and summaries included in the articles, the numbers of "embryo transfers" are often found under "outcome", where they signify successful laboratory fertilizations.

The transformation of the invasive aspects of treatment to the status of preparatory work or measure of success is most evident where "conventional" IVF is extended with the micromanipulation techniques such as intracytoplasmatic sperm injection. It is mirrored in titles like: "Microinjection of Human Oocytes: A Technique for Severe Oligoasthenotheratozoospermia,"[220] or "Routine Application of Partial Zona Dissection for Male Factor Infertility."[221] Such titles exemplify an exclusive focus on fertilization and laboratory techniques in defining or demarcating 'treatment' for male infertility. The abstract preceding the former paper, for example, includes only the following under 'interventions': "Sperm was injected subzonally or directly into the ooplasm". This statement, again, produces a sense in which the whole range of bodily interventions implied in this technique, does not count as 'intervention' in its own right. This conceptualization is entirely consistent with the displacement of the problem from bodies to Petri dishes in laboratories and its definition as a failure of fertilization in vitro. But to the extent that it actually concerns a reconceptualization of bodily intervention as such, it reduces the measure in which intervening in female bodies as a prerequisite to such conceptualizations remains visible.

In this section, I have described several instances of a pattern in which the interventions on female bodies involved in the construction of fetuses and couples as patients are erased to a considerable extent. All examples discussed involved the reconceptualization of intervention as such. My analysis of the medical work reported shows that the meaning of intervention changed toward prevention, diagnosis, preparatory work, or even measure of success. All these transformations have in common that they pull attention away from the interventional character of the events thus relabeled. This pattern of deletion has the cumulative effect of reducing the visibility of the many required interventions in a body that is not coextensive with the couples and fetuses designated as the 'patients'.

3. THE PURIFICATION PATTERN[222]

A second way in which the visibility of the hybrid nature of the new patients is reduced is to be understood as "purification". Whereas the deletion pattern downplays the *artefactual* nature of 'fetuses' and couples' as patients, that is, the conditionality of their existence upon technical-medical interventions and work directed at female bodies, the purification pattern addresses the second sense in which the new patients form hybrid entities. Beside being mixtures of nature and technology, of bodies and machines, 'couples' and 'fetuses' constitute also hybrid forms of *individuality*. The pattern I describe in this section results in the "purification", or resolution of the ambiguity of their messy in-between status of being neither one nor two, or both. The purification consists of the discursive establishment of a firm connection between fetuses and couples on the one hand, and subjects whose individuality is *un*ambiguous on the other. The ambiguities are resolved by relating couples and fetuses to what, in the end, appear to be seen as the true individuals concerned.

As it turns out however - and this shows the political nature of any such purification process - it is not women who are thus named as the 'real individuals' concerned. Despite the crucial role of women and their bodies in the technological procedures, these practices are in numerous instances conceived of as involving, above all, *men and children*, rather than women. The silence of these texts on the relationship between female bodies and fetuses and couples is paired with a relative overemphasizing of the connections between the new patients and the subjects of children and men. These connections are established by association, analogy, equation, or juxtaposition, and have the effect of transferring some of men's and children's unambiguous individuality onto the new patients. Unavoidably, however, highlighting these connections means that the relationship of fetuses and couples to women as embodied individuals recedes further into the background. If these practices are, actually and ultimately, still conceived of as being primarily concerned with men and children, the shift of medical problems between individuals (male and female, future and present) becomes yet more difficult to recognize.

In what follows I will explain the mechanisms involved in the purification pattern by focusing on *distributions agency and properties*. As I will show, these distributions play a crucial role in determining what and who can become construed as an individual in these texts. Thus I will try to make clear how these discourses establish a suggestion of being more about men and children, and their lives, capacities, futures and functioning, than about the women through whose bodies all this becomes realized.

3.1 From Couples to Males

The attribution of male infertility to a couple has its basis in transformations and relocalizations of the problem. In the context of IVF, male infertility becomes a pathology that does not occur in a male body, but instead in a Petri dish: there, the interaction between gametes becomes identified as the locus of trouble in male infertility. From this phenomenon of 'gamete interaction', then, 'the couple' emerges as a singular patient. This 'couple' may be infertile due to male or female 'factors': the infertility is a property of the couple as a singular entity, in which subsequently parts, factors and features are distinguished as female or male. Whereas gamete interaction as a singular phenomenon becomes the locus of pathology of the couple, an observable process, and an object amenable to treatment, the concept of 'interaction' implies that the gametes (sperm and oocytes) are still discernable as separate, male and female, entities. This distinction between the two entities enables a description of their 'interaction' in which properties and agency are distributed in particular ways. Via the distribution of the properties and agency of these male and female micro-entities, a particular discursive connection is established with bodies and persons.

However, the remarkable thing is that the distributions of agency and properties within 'gamete interaction' - despite the reciprocity implied in the term 'interaction' - is in fact thoroughly asymmetrical. Even if one considers male infertility a fertilization problem (instead of a spermatogenesis or sperm quality problem), this alone would still allow a location of the origin of the problem in a male "part" of the couple. However, in the Petri dish the problem can subtly slip from being a property of the sperm cells to becoming a property of the oocyte, as is exemplified in the following description from a paper on micro-fertilization techniques:

> Union of gametes either in vivo or in vitro requires penetration of the *investments* surrounding the oocyte and *hinder* sperm penetration. These *obstacles* are the cumulus oophorus and the zona pellucida. In oligospermia, the statistical likelihood that a sufficient number of *cells capable of traversing these barriers* is reduced, and even though some normal sperm are released, the male is functionally infertile. ... In recent years, the technology of micromanipulation has significantly improved, and it is now possible to *circumvent the barriers* to sperm penetration and reduce the number of sperm cells needed to achieve fertilization.[223]

In this quote, "functionally infertile males" and reduced capabilities attributed to sperm cells are mentioned, but the description of the anatomy of the oocyte tells another story. Investments "hindering" penetration, "obstacles," and "barriers" conjure up an image of the (normal) oocyte as a troublemaker. The micromanipulation techniques employed are all designed and conceived of as strategies to circumvent the "barriers", and as attempts to give spermatozoa "opportunities" that are withheld by the oocyte. In accordance with the location of the problem with the oocyte, it is the oocyte that is seen as needing "treatment", and as the "target" of manipulations. In another paper on one of these techniques, the authors write:

> As a result of improved technology, gamete micromanipulation has gained great attention in recent years. Micromanipulation techniques *circumvent the physical barriers* to sperm penetration, especially the zona pellucida, and the number of sperm needed to achieve fertilization may be greatly reduced. Several micromanipulation techniques have been described. The least invasive technique, involving creation of a hole or incision in the zona pellucida to promote sperm penetration, is referred to as "zona drilling", "zona cutting" or "pzd" [partial zona dissection].[224]

Since the problem lies in the oocyte's being "obstructive", it should be no surprise, that, when all these manipulations do yield a fertilized oocyte, this result is described in terms of the fertilizing capacities of the sperm cell. Once the barriers are removed, the sperm cell is given the opportunity to realize the potential it had all along:

> The partial zona dissection and subzonal sperm injection techniques have in common that *spermatozoa, whose chances* of penetrating the zona pellucida would otherwise have been remote, *are given the opportunity to bind* to the oolemma. Spermatozoa from men with severe teratozoospermia (</= 5% normal forms) *can fertilize* one fifth of all *oocytes treated* with partial zona dissection. Such embryo's though rarely implant. A recent study showed that 16% of

spermatozoa from severely teratozoospermic men *are able to form pronuclei when inserted into the perivitelline space.*[225]

This rendering of sperm cells as the active agents in fertilization, with the oocytes mere passive entities to be acted upon, strongly resembles the physiological accounts of fertilization analyzed by, among others, Martin (1992) and Pfeffer (1987). The inscription of male and female cells with stereotypical gender roles and behavior has cultural and historical roots as deep as Aristotelian accounts of the male as the form-bestowing active principle and the female as the passive, receiving formless matter and nourishment. Even though in recent years such accounts have been thoroughly revised, its stereotypical distribution of agency and properties can be seen to linger on, not only in the popular imagination about this 'natural' process, but in these state-of-the-art high-tech renderings as well. Apparently, this pattern is so difficult to shed that it is even maintained in the context of male *in*fertility, that is, in the case of immotile, misshapen, or otherwise defective sperm cells.[226] As one researcher commented:

> Reproductive biologists tend to have one-track minds. We conceptualize the sperm as a well differentiated missile and the egg as a sessile stationary target. It is difficult for us to conceive that sperm and egg could be brought together by techniques other than sperm microinjection."[227]

The obvious military metaphors here are interesting enough, with their implication made clear that therapeutical possibilities have been thought of in accordance with the two main possible ways in which a war can be fought: defense or attack.[228] The possible consequentiality of this is suggested in the following quote from Van Rijn- van Tongeren's analysis of the function of metaphors in medical texts:

> Therapies are linked with theories and metaphors constituting medical theories thus determine the therapeutical possibilities. When therapies are considered inadequate, alternative theory constituting metaphors have to be found. The aspect of highlighting and hiding is important in connection with medical theories, as valuable therapeutical possibilities may be hidden by the metaphors constituting those theories. Analysis of the way in which the donorfield of a metaphor is structured by the donorfield may reveal which aspects of a phenomenon are highlighted and which are hidden and thus help to find alternative metaphors to establish new theories.[229]

However, it is above all the attribution of agency in fertilization (and its perceived causative link to technological development) that is of interest here. In male infertility the sperm cells, by definition, do *not* play their expected part in the 'gamete interaction', even though they are placed near the oocyte in as "favorable conditions" as possible. And yet there is, according to such descriptions, nothing really wrong with them. The fertilization failure is not due to intrinsic properties of the sperm cells, which is proven by the techniques that merely assist, and clear away the barriers. The act of fertilizing is still squarely attributed to intrinsic properties and activity of the sperm cell. There is, according to such depictions, it seems, no form of male infertility so bad that it would actually constitute a dysfunction attributable to the male reproductive system: it is literally "irrespective of the quality of the semen" that sperm cells are "able to form pronuclei". This way one can see how, within the language on micro-level male reproductive dysfunction, a certain lack of problematization of the male reproductive body is still at work.[230] The redistributions of agency and properties on the level of cells described above are paralleled by similar patterns on the level of bodies and persons. The paper "In Vitro Fertilization Techniques with Frozen-Thawed Sperm: A Method for Preserving the Progenitive Potential of Hodgkin Patients" can serve as an illustration here.[231] While it was the female "contribution" (in the form of everything required to make large numbers of oocytes available in the IVF laboratory; in this particular case a total of 110 oocytes from five women), that was magnified to the extreme, while the male contribution in terms of sperm cells remained "extremely impaired", the authors consistently phrase the outcome of their proceedings in terms of progenitive capacities of the males:[232]

> Hodgkin patients often suffer from oligospermia, even before starting their therapy and in only about one in four Hodgkin patients is the semen suitable for cryopreservation. In this paper, we present an approach to *preserve the patients' reproductive capacity* by combining sperm banking and in vitro fertilization (IVF). ... We would like to recommend that *every patient* undergoing a cancer therapy that eventually could cause gonadal damage must be encouraged to cryopreserve his semen before starting his treatment, *whatever the quality of the semen. This semen can fertilize* in vitro after thawing, even in case of grossly impaired sperm parameters. ... The results of IVF-ET and related procedures seem very promising for the *maintenance of the progenitive capacity of Hodgkin patients.*[233]

Thus the distribution of capacities and agency between male and female entities at the microscopic level of cells and gametes is paralleled at the macroscopic level of bodies and persons, where a similar reversal is accomplished. No matter how infertile, it seems, a male body is never really infertile, provided all 'external conditions' (other bodies, that is) are made conducive to letting this body function.

The point here is, that this is indeed what these accounts seem to relate: whatever is required in the way of manipulation of "the couple's body", in the end it all amounts to having made a *male* body function properly. This body may have remained entirely untouched by technology, and yet this technology is seen as having enabled this body to use *its* capacities. This is how duplicity of 'the couple' is, partially at least, erased. By repeatedly connecting the treatment of the couple to restoration of function and establishing capacities of the male body, this association acquires an emphasis over the material and practical connection of 'the couple' to the female body. Because, ultimately, the technology is about enabling male individuals to realize their own potential, the shift of the problem and its treatment between individual bodies and even the very involvement of women have become all the more difficult to recognize.

3.2 From Fetuses to Children

I will shift focus once more to fetal surgery.[234] As described in the previous chapter, we find here a continuous effort to distinguish between the fetal patient and the rest of the female body. The adjective 'maternal' preceding a physiological process or body part, in general, is the accepted way in perinatal medical discourse to refer to women's physiology or anatomy. This distinction between "mothers" and "maternal factors" on the one hand and "fetuses" and "fetal factors" on the other is carried through in fetal surgery to the point of becoming absolute. Just as interventions on "the mother" are thus distinguished from interventions on the fetus, so too are effects of the procedures on "the mother", if represented, distinghuished from effects on the fetus. By implication, interventions and effects on fetuses are *not* interventions on women, and vice versa. Thus, while the "mother" may be given anaesthesia, it is not she, but "the fetus" who undergoes "shunt placement". Conversely, while the fetus may deteriorate or die after/from the procedure, or the preterm labor induced by it, the 'mother' may be said to suffer no complications or morbidity. But this distinction does not mean that the 'fetal' and 'maternal' parts are given comparable weight. Rather, the distinction is effectuated in large part by attributing most properties, characteristics, or events to the fetus, while being silent on the woman. Or, more precisely, in ascribing events and characterizations to the fetus, they are, by implication no longer belonging to the woman. Pregnancy itself has become primarily a process a fetus goes through, or even, acts out, and only in rare, disparate instances recognizable as a process experienced by a woman, or as a capacity of her body.

This pattern of distributing agency and properties between 'maternal' and 'fetal' parts of the patient, when combined with the strong emphasis on the developmental potential of the fetus in terms of what it might become, that is, the future child, produces a particular effect. An effect analogous to the one described in the previous section on 'the couple': through association, comparison, analogy, juxtaposition and even equation, a strong connection between the fetus and the future child is established. Its connection with the future child is often stressed *over* its connection with a woman. Thus, the purification pattern reduces the inherent ambiguity of the fetus' status as an individual patient by emphasizing its link to a prospective child over its actual embeddedness in the body of a present individual.

An instance of the pattern is formed by the way the duration of pregnancy, a matter that could as well concern the female individual, becomes "gestational age", then age of a fetus, and finally, in retrojective equation, the "age" of a child:

> Only four of the children that entered the study had polyhydramnios, although all of these children (61) had an ultrasound diagnosis made before 24 to 25 weeks.[235]

> Most of these babies don't have polyhydramnios at 24 weeks so we can't actually use it as a risk factor at that age.[236]

By the same token, having an abnormally large amount of amniotic fluid (polyhydramnios) is a property of babies and children, while agency can be seen to shift entirely from women to uncanny self-reliant (unborn) children entering studies on fetal surgery and having ultrasounds made.

The developmental perspective plays a constitutive role in the construction of the fetus as patient. This perspective produces a distribution of properties and agency that stands in sharp contrast to the agency and activity attributed to the male body in fertilization, as described in the previous section. Reminiscent of the same Aristotelian scheme of passive, feminine matter acted upon, the growing, developing and generation of a new individual child is not something that is attributed to capacites and activities of (parts of) the female body, but an activity of an entity distinct from that body. By focusing on the (possible) result of this process, an individual child, the characteristic of *being* an individual in a non-ambiguous sense is, by analogy, transferred onto the fetus. In the quotes above (extracted from a discussion between Harrison and Arensman of the results published by the first) one can see how, after the pregnancies are over, the story of the interventions during pregnancy is retold in a way that retrojectively has 'children' rather than 'women' acting in it as the subjects concerned. As a result the link of fetal patients with children is established and highlighted over their connection with women.

This connection between fetuses and children is an *ontological* one because it concerns the construction of a relationship between fetus and child of actually *being* the same entity. Fetus and child are essentialy the same, their differences mere contingencies of time and space: the fetus *is* the young child that is temporarily located elsewhere.

The close interrelationship between the emphasis on development, the equation of fetuses with young children, and the mentioning of children (instead of women) as the individuals receiving the medical care involved in prenatal intervention, is evident in the following quotes as well:

> The physician's ability to *extend medical care to the very young* has rapidly evolved over the last decade. Even though fetal therapy is still developing, it seems clear that when appropriately utilized, with specific protocols and institutional approval such therapy is valuable.[237]

> *Some fetuses* have massive ventricular enlargement when first encountered. Others have equivocal findings and may or may not show progression of the ventricular enlargement. *Some individuals* will have relatively enlarged ventricles at the first examination and then either return to normal size or continue to be abnormally large but decline parallel with the normal zone. In early infancy *some of these babies* appear to be developing normally, but the long-term prognosis in terms of intellectual development for *these children* is unknown. ... Two of *the survivors* are showing nearly normal development. *One had his shunt placed at 23 weeks' gestation,* but an amniotic fluid leak was noted a few days later. *He was delivered* at 27 weeks when it was thought that amnionitis was developing. His shunt was in place and functioning at birth and was replaced with a ventriculoperitoneal shunt. Whether the 3 weeks of intrauterine shunt treatment contributed to *his outcome* is uncertain. *The last child* had his shunt placed at 24 weeks' gestation. At 27 weeks his ventricles enlarged and the shunt could not be seen on ultrasound. *He had a second shunt inserted* at that time. *He was delivered* at 32 weeks' gestation because of spontaneous rupture of membranes, and his shunt was again out of his head.[238]

These quotes indicate a clear pattern. While the fetus's status as individual patient is ambiguous because of its intrinsic embeddedness in an (other) individual, this ambiguity dissappears to a certain extent. The dual possibility of simultaneously counting as one (pregnant individual) and as two (woman and fetus), is reduced to just one, by discursively establishing an equation between the fetus and an unproblematic, unambiguous individual. However, this resolution of the duality of woman and fetus is accomplished, not by reconfirming the individuality of the pregnant woman, but by replacing hers with that of the (future) child. The whole pregnancy, and everything that occured within it, is described as having happened to one individual, a child, and, to that extent, as never having involved a woman at all.

Significantly, one of the leading practitioners of fetal surgery, M.R. Harrison, adorned several of his publications with the following quote from Samuel Taylor Coleridge:

> The history of man for the nine months preceding his birth would, probably, be far more interesting and contain events of greater moment than all the threescore and ten years that follow it.[239]

In its literary eloquence, it contains all the elements of the purification pattern identified in this section, including ontological continuity between fetus and child, pregnancy as experience of the future person - an experience, moreover, of an importance never equalled again in this person's entire later life, and the absence of any woman.

4. EVALUATIVE COMPARISONS, OR: HOW TO RENDER INCOMPARABLE THINGS COMPARABLE

One of the places where one, naively perhaps, would have least expected the implication of female bodies in these practices to be discounted, is the explicit discussion of the advantages of the proposed treatments. Nevertheless, it is precisely in some of the evaluations of the technologies, presented as scientific rationales for the procedures, that one can see the deletion and purification mechanisms at work. To add some more relief to these patterns, this section describes how they can be recognized in specific arguments, evaluations, and comments on the pros and cons of the new technologies. It will be shown how such evaluations can be set up in ways that effectively *add* to the erasure of interventions in female bodies from the accounts, and the deconstruction of individuality of women in order to stabilize that of others.

That the advantages of the new technological procedures are not self-evident, even to the practitioners themselves, is something that is easily derived from numerous statements throughout the literature. Within the medical community, fetal surgery is still highly controversial, and for years, a consensus concerning the question whether IVF can or cannot be considered a worthwhile technique for male infertility was lacking.[240] Low success rates in IVF in general are frequently attributed to low sperm quality. It is quite interesting to see the matter-of-fact-like tone in which statements concerning the association of low sperm quality with poor success rates in IVF are made:

> De Kretser and co-workers have reported that when there is a combination of three or more defects in the semen analysis, fertilization in vitro diminishes to <8%.[241]

Over the past decade in vitro fertilization (IVF) has become a routine and accepted tool in the armamentarium of infertility treatment modalities. It has become clear that while the technique is capable of providing a succesful solution for certain diseases, it has major limitations in solving other fertility problems. Male infertility has been one of these fields, in which a small fraction of oligospermic men have benefited from IVF, whereas a large fraction of oligospermic males were refractory to the technique.[242]

The message in these quotes seems quite straightforward; IVF was a useful technique, *except* in cases of male disorder. If this is the case, then why was IVF so widely used as a treatment for male infertility, and so often described as the most promising direction in the search for treatment of male fertility problems? One way to explore this question is to examine what scientific claims about the utility of IVF for male fertility problems look like.

The key to the puzzle lies in the fact that usefulness or success of a medical technique itself is no straightforward matter. Success is always a relative issue, a judgment made in comparison with something else. It is the choice of what to compare it to that makes the difference in IVF for male infertility.

In the quotes above, the (implicit) comparison is between success rates of IVF with normal and with defective sperm, that is, the comparison is between IVF for female and for male problems. Even though IVF for female problems has notoriously low success rates, it is considered a "routine" treatment, the success rates of which have become a norm against which IVF for males is judged. Set against this norm, IVF with defective sperm is apparently falling short. The flip side of this kind of reasoning is that when and where success rates with defective sperm are improved to the point of being comparable to those in cases of female disorder, that in itself can become considered justification enough to include male infertility cases.

> In our patient sample, the application of IVF for purely male infertility has increased from 3.2% in 1981 to 27% now, *as a consequence of* the improved fertilization rates. In our IVF programme, *only slightly lower pregnancy rates per embryo transfer* are observed in the patient sample, *justifying* inclusion of these [male factor] patients.[243]

These results are, however, an exception. The general picture is that of diminished or ambiguous results for IVF in male cases. Nevertheless, there are other ways to establish the claim that trying an elaborate procedure such as IVF is worthwhile. One way is exemplified in the following quotes in which a different evaluative comparison is made:

> Subfertile men treated with *conventional methods*, including medication, varicocele embolization and/or intra-cervical artificial

insemination, have a probability of conception of *3.3% per cycle*, ...
With GIFT the probability of conception is *18% per cycle* ... The
succes rate of IVF is *8% per cycle* ....[244]

It appears that the use of abnormal semen decreases the pregnancy
rate of GIFT, but not of IVF-ET. *Nevertheless*, the ongoing pregnancy
rates of *18% per cycle* with GIFT and *8% per cycle* with IVF are
higher than those atained with conventional methods of treatment of
male infertility. Indeed, the probability of conception after treatment
of varicocele is ca.4% per cycle and the average succesrate including
all types of conventional treatments of male infertility is ca.3% per
cycle.[245]

Here we find "*conventional methods*" as the basis of comparison and
evaluation of the use of IVF and the related technique of GIFT[246]. This
concept of conventional treatments includes hormonal or surgical treatments
of the male, as well as artificial insemination (AI) and intrauterine
insemination (IUI). The inclusion of these latter two procedures confirms
the observation that shifting treatment of male problems to the female body
is not considered a significant change of approach. Rather, it is the transition
from "conventional" to (probably) experimental approaches that is set up as
the main contrast on which evaluations are based. Though the suggested
procedures (IVF and GIFT) are recognized as "invasive", and should,
therefore, be tried only as a last resort, the question of who is to be
"invaded", for whose problem, does not arise.

Success rates of 8% per IVF-cycle and 18% per GIFT-cycle, such as
mentioned in the above quotes, in themselves probably do not strike an
interested outsider as particularly high.[247] Nevertheless, the authors arrive at
the conclusion that their approach is very promising, and encourage their
colleagues to follow suit ("should be attempted"). It is mainly through the
repeated juxtapositioning of their results against the even worse prospects of
"conventional treatment" that their claims become convincing.

There is more to be said, however, about the means by which IVF, and
fertility treatments in general, are rendered comparable. Since scientific
judgments about the clinical use of IVF, including the two types mentioned,
rest in large part on such comparisons, this is no trivial point. Comparisons
between IVF in male cases and other treatments are made possible by *calculating results per cycle*. This is not restricted to IVF alone, but a general
convention in the medical discourse on fertility treatments. The quotes above
are good examples of this convention.

The notion of a cycle derives from the monthly menstrual cycle of
women. Within the context of fertility treatment this 'cycle' has undergone an
interesting transformation. From being a physical process occurring in a

female body it has come to denote a much more abstract kind of unit.[248] It now represents one opportunity for impregnation, irrespective of the way this impregnation is made to come about, or the measure of manipulation and intervention in the cycle.

"A cycle is a cycle is a cycle." However, the 'cycle' in conventional treatments of the male, for example, varicolectomy,[249] denotes the temporal unity of the monthly cycle in the female body. Once the surgery on the man is done, no further medical interference or surveillance is required. Consequently, the difference between one and, say, eight cycles is only a matter of a longer period of "trying" having passed.

A GIFT or IVF cycle, on the other hand, denotes one trial of the treatment itself. In the case of GIFT, for example, this cycle involves a month of daily administration of high doses of hormones, constant monitoring of hormonal levels and follicle development with variously invasive diagnostic techniques, an oocyte aspiration and, if all went well, the placement of gametes in the fallopian tube under general anaesthesia. Here the difference between one and eight cycles gets a somewhat different meaning: while one cycle is burdensome but probably doable, it borders on the unimaginable what it would mean to have a cumulation of eight such 'cycles'.

Any calculation requires formalization that, by definition, implies making a distinction between relevant and irrelevant aspects and differences. In calculating results per cycle to make such incomparable treatments as varicolectomy and GIFT comparable, abstraction from the material aspects of the treatments is required. Thus, in this case, differences in number and nature of physical interventions have to be discounted. In addition, the formalization of the results of treatments of men in terms of cycles, including those "conventional" treatments directed at the male body, requires discounting the differential distribution of medical interventions between the sexes. The concept of a cycle is thus not only dissociated from its meaning as a physiological process, but as a specifically female process as well.

Turning our attention to fetal surgery now, we find evaluative comparisons showing similar choices of what constitute relevant aspects to be foregrounded and contrasted, and, consequently, similar choices of what can recede into the background. One of the most central criteria for deciding whether fetal surgery is indicated, as they were formulated by the International Fetal Surgery Society, concerns the availability of effective postnatal treatments.[250] This seems to acknowledge the relevance of the difference between prenatal and postnatal treatment with respect to required measure of intervention in female bodies.

However, in reviewing rationales given for fetal surgery, it becomes clear that such acknowledgment is not necessarily always the case. It is quite possible that reference to the way a choice for prenatal intervention would affect the female body is actually absent from the discussion. When the utility of fetal surgery is discussed, possible advantages are commonly identified by comparison with operating on *born* children. However, the advantages thus identified often are premised on discounting the fact that operating on the fetal body, unlike operating on the child's body, always implies operating on the female body. The contrast set up is that between fetuses and children, not between pregnant women ("women-plus-fetuses") and children. Consider the following table, taken from an article on animal experiments for prenatal surgery of cleft lips:

Table I Advantages of in Utero Surgery
1. Prevents irreversible vital organ damage
2. Allows organ system redifferentiation and regeneration.
3. Unique wound healing milieu (no scars)
4. Immature fetal immune surveillance
5. No required extracorporeal support
6. Transplacental administration of drugs by means of maternal circulation
7. Infection can be combatted using maternal immune factors.[251]

In tables such as these, the implicit comparison is that between operating on fetuses and born children, by discounting the implicated interventions on female bodies. One can see how, if fetal surgery would be conceived of as surgery involving women, some of these "advantages" would disappear immediately: the woman *will* require "extracorporeal support" (see 5), and her risks of infection will have to be "combatted" with antibiotics (7). Especially the point about fetal wound healing (3) is revealing.

Many scientists are, understandably, fascinated by the fact that, during a specific period in gestation, fetal tissue heals very rapidly, without forming scar tissue. Presently, many research efforts are directed at exploring this phenomenon.[252] In fact, this line of research exemplifies a concrete instance of the way clinical therapeutic experiments and basic scientific research intertwine both practically and politically, rather than being a one-way street from basic research to clinical application. As Casper (1995) described, it were the surgical experiments on pregnant women that called attention to the phenomenon of fetal wound healing that subsequently became the object of study in clinical, animal, and in vitro models. This scientific 'spin-off' then, bestows some legitimacy on the surgical experiments, since it thus helps science forward, while on the other hand, this way of achieving scientific progress is partly legitimized by its association with a therapeutic setting and claims of further expected therapeutical benefits. This particular research object strikingly shows how scientific rationales for clinical surgical experimentation tended to be produced after and through clinical intervention rather than the other way around.

The main point here, however, is that the presentation of fetal wound healing as a clinical rationale for intervention, as in the above quote, cannot but rest on exclusion of female bodies from consideration. Sometimes, the scientific fascination whith this phenomenon can slip into enthusiastic phantasies about the prospect of 'preventing' surgical scars on children by corrective fetal surgery, even though reaching this goal would mean cutting open a woman's belly twice - delivery after open fetal surgery (hysterotomy), as a rule requires Cesarean section, to prevent rupture of the relatively fresh surgical wounds on the womb during labor - without, obviously, her having the benefit of any miraculous capacity for wound healing. Consider, for instance, the following argument, taken from the same article on cleft lips quoted before:

> Reconstruction of facial clefts can restore proper anatomic form, but some disfigurement inevitably remains, since formation of a scar is a normal physiologic sequela of skin wound healing. The hypothesis that fetal repair of a cleft lip or a cleft palate might ameliorate the dysmorphogenesis as intrauterine growth then proceeds or even eliminates facial scarring is not a new concept. The objective of this presentation is to demonstrate that in utero cleft lip repair is technically feasible in the A/J mouse and that the period of gestation is not prematurely interupted by such intervention. In addition, the concept of an absence of scar formation in primary fetal wound healing is confirmed.[253]

The article concludes by saying that "the absence of a lip scar after human cheiloplasty [the postnatal surgical repair of cleft lips] may require the as yet undefined advantages of fetal wound healing."[254] A similar way to identify "advantages" of fetal surgery is exemplified in another article:

> But there are other advantages to fetal surgery that have even greater importance. They are as follows: (1) The potential for healing and regeneration are many times greater in the *fetus* than in the *fully formed* [sic] *infant*. Surgical wounds would heal rapidly - some without suture. ... Thus, in many ways, *fetal surgery may become safer than operating on the infant.*[255]

The qualification of fetal surgery as 'safer' can only mean that women are left out of the comparison; for them, such an evaluation would in all cases turn out the opposite.

There is no easy, self-evident way of knowing how to weigh the advantages of fetal surgery, and what to compare it to, in order to evaluate its risks and benefits. Indeed, any choice of what to compare any intervention to, in order to establish a 'medical justification', in a way, carries a bias, for the act of constructing such an analogy necessarily involves making a distiction between relevant and irrelevant aspects and differences, making a distinction between what counts as foreground and what as background. So, too, the choice encountered here carries a bias. Rendering such different procedures as open fetal surgery on a pregnant woman and pediatric surgery on a child or infant comparable, necessarily involves making distinctions and choices between relevant and irrelevant differences. In the examples given above, it can be seen how easily the difference in measure of intervention in the female body can become relegated to the irrelevant background, when the contrast foregrounded is that between a fetal patient (instead of a pregnant patient) and a pediatric or neonatal patient. This reduction of the visibility of the implied intervention on female bodies produces a magnification of the perceived advantages of fetal surgery.

Thus, many of the evaluations of the technologies can be seen to constitute instances of both the discursive mechanism of 'deletion' and that of 'purification'. Often, these evaluations are based on comparisons with other treatments that differ precisely in measure of intervention in female bodies. By implication, the interventions are deleted from the comparison, and relegated to the background irrelevant to the evaluation.

Moreover, the evaluations contain elements that can be described as purification, because they highlight connections between hybrid couples and fetuses on the one hand, and children and men on the other. This reestablishes a sense of the patient's individuality, at the direct expense, however, of the recognizability of the involvement of women and their bodies. In setting up analogues to fetal surgery and in vitro fertilization, to argue their efficacy and advantageousness, treatments for 'fetuses' are usually juxtaposed to treatments for children, while treatments for 'couples' are compared to "conventional" treatments for men. The invocation of the images of these more usual 'patients' as relevant analogues, makes individual patienthood for 'couples' and 'fetuses' seem less ambiguous and more acceptable.

5. THE CO-PRODUCTION OF TECHNOLOGICAL PRACTICES AND LEGITIMACY

The patterns of deletion and purification accomplish a reversal that considerably enhances the appearance of rationality of these technological practices. By 'deleting' the interventions on female bodies involved, the constructive medical work that generates fetuses and couples as patients suffering from particular problems is erased and, with it, the artefactual nature of these 'patients'. Once they have acquired 'natural' status in this way, the process in which they are constructed can then be presented, in reverse, as a consequence of a prior existence. The interventions providing the material conditions for fetuses and couples to emerge as 'patients' in the first place, are perceived post hoc as following from the 'medical facts' that it is actually couples and fetuses that suffer from medical problems and are in need of treatment. Thus, the *result* of intervention, that is, the technological procedures performed on women's bodies through which congenital disease and male infertility are transformed into a fetus's and a couple's medical problem, is changed into the *reason* for intervention in women's bodies. The inconspicuousness of this reversal is to a great extent enhanced by hardly ever presenting interventions as interventions on female bodies, and hardly ever having women figure as individual patients. It would be much harder to present open uterine surgery or all the other invasive procedures performed on women's bodies as treatments of 'others', if women would still occupy the position of the central, individual patient, and be referred to as such.

Instead, as we have seen, there is an additional tendency to further downplay the constitutive connection between fetuses and couples and women's bodies. When fetuses and couples are linked to more conventional types of patients and their medical problems, it is to children and men, rather than women. Thus, when the relationship between the couple and the male individual, and between the fetus and the future child is highlighted over their relationship with female bodies, the new patients' "in between" character of being neither one nor two, or both, is, on a discursive level, resolved as well. By systematic association, analogy, comparison, juxtaposition and even equation, treating couples and fetuses is presented as being merely a specific, new form of treating men and children. With IVF, the progenitive capacities of men are restored, and fetal surgery is a form of early surgery on the child.

From the partial perspective which considers women as the central patient, and which regards their individuality and physical well-being as issues that matter, individual women are seen walking in and out of waiting and treatment rooms, climbing on and off gynaecological chairs, letting themselves be monitored, medicated and operated upon. From the equally partial perspective of medical technoscience, these practices constitute events in which hardly anything seems to happen to women as such. Instead, we are told that what is going on here, is about couples and fetuses, and, if about clear-cut individuals at all, it is about men and children. 'Couples' are treated to enable men to function reproductively; 'fetuses' are treated to promote children's health. Thus, in more than one sense, women are not the issue.

This way, questions of legitimacy of these technologies (considered, for instance, in the light of women's health or a sound reproductive politics) are, right at the onset of the development of these scientific and technological practices, to a large extent prevented from arising in this context. Or, perhaps it is the other way around, legitimacy being actively produced: potential problematic aspects of medical interests of "in between" entities are kept at bay, through their conceptualization as being merely forms of culturally accepted medical interests of unproblematic individuals. The ultimate justification of these technologies implicitly refers to undeniably legitimate interests: that of men to become father, and that of children to be as healthy and have as much opportunities in life as possible. No one in their right mind (and, one might add, least of all the caring wives and mothers) would contest the legitimacy of these interests. Therefore, the stronger the chain of translations leading from these unproblematic interests and individuals to fetuses and couples, fetal surgery and IVF, the more difficult it is to contest the legitimacy of these new technological practices, and the concomitant shift of responsibility, risk, and suffering onto women. In this sense it can be argued that the discursive patterns of deletion and purification, and the resulting accounts of what IVF and fetal surgery are 'about', achieve justificatory as much as descriptive ends.

CHAPTER 4

ELUSIVE BODY BOUNDARIES AND INDIVIDUALITY

"Of course, who controls the interpretation of bodily boundaries in medical hermeneutics is a major feminist issue."[256]

1. INTRODUCTION.

In the previous chapters I described a variety of mechanisms which have contribued to the emergence of fetuses and couples as patients and which have in part facilitated the development of IVF and fetal surgery. Arguments were presented to show that medical interventions redefine and relocate the problems addressed in ways that simultaneously transform the patients suffering from these medical problems. Precisely in doing all the medical work necessary to arrive at the particular diagnoses involved in fetal surgery and IVF for infertile men, new problem-definitions as well as new patients have been construed. Prior to artificial fertilization and modern prenatal technology, couples nor fetuses were considered to be patients as such.

Technologies, of course, do not develop in a social-historical vacuum. They arise from and subsequently become part of particular historical contexts of action that, in turn, constitute a set of pre-existing conditions and facts for further developments. One of the pre-given facts in reproductive technologies, the consequentiality of which can hardly be overlooked, is the centrality of gynecology and its traditional object, the female reproductive body. Both in vitro fertilization and prenatal technologies evolved out of intensified exploration of the female reproductive body, yielding new interventional methods for female conditions. Thus a framework was created in which the reproductive couple as well as the fetus gained visibility and were made into newly demarcated objects of treatment. It takes work on the female body to exteriorize ova and produce the 'interface' of gamete inter-action so crucial for the construction of the couple as the patient in male infertility. Similarly, it takes work on the female body to generate the images and data that make up the congenitally affected fetus. This focus on work and action (in the form of medical intervention in bodies) rather than knowledge (of how bodies function) provides better insight in the mutually constitutive relationship between the body and the development of technologies. Far from being the result of the application of existing knowledge, technologies tend to be used on bodies while it is still highly disputable whether the relevant medical knowledge is available. This knowledge, specifically that regarding fertilization processes and fetal development, is clearly what results from these practices.

The medical-technical work of the past two decades has caused significant shifts in perspective and focus. We have ended up with gynaecological practices, some of which have the explicit purpose to help others than this field's traditional patient: women. In vitro fertilization has become, among other things, a method to restore or maintain functions of the male body; fetal surgery is geared toward curing "children". Corresponding to these shifts are the medical disciplines now in cooperation with gynecology in the two technological practices. Both IVF and fetal surgery are interdisciplinary practices involving teams of specialists from multiple fields. Besides gynecological knowledge and skills, andrological ones are involved in IVF for male conditions, while neonatologists and pediatric surgeons are prominently present in fetal surgery. Boundary crossings between disciplines, and the resulting hybrid medical fields of IVF and fetal surgery, form the institutional correlates to the new hybrid patients these practices have generated. Where couples appear, gynecologists need andrological knowledge; where neonatologists and paediatric surgeons cooperate with gynecologists, fetuses become the focus. The dissolution of boundaries between traditionally separated fields of work corresponds to a comparable dissolution of boundaries between their respective objects/patients, that is men and women, and women and children.

But in all these changes mergers and boundary crossings, a huge paradox has become evident. This paradox stems from the fact that despite all these changes one thing has remained stable. While it is true that most learning is achieved by doing and making mistakes, the strange thing here is that while the 'doing' still concerns female bodies, the 'learning' is about helping men and children. Whereas both patients and problem-definitions have been radically transformed, the primary objects of intervention have, by and large, remained the same. It is still ovaria, uteruses, vaginas, tubes, in short, the female reproductive body that undergoes most of the interventions and manipulations involved in both practices. Paradoxically, however, it is precisely women who seem to have become unrepresentable in these new practices. This situation appears all the more puzzling if one considers its counterpart: whereas the other two patient categories, men and children, are, each in a different sense, absent as objects of clinical interventions, they are represented as individuals. As I showed in the previous chapter, when fetuses and couples are treated, women are not around, but children and men all the more so. While it is still women's bodies on which the various specialists now jointly work, and not those of men or children, the involvement of women's bodies evaporates from both discourses. These bodies are no longer visible as discrete entities, or considered as the body of an individual human person.

2. BODY BOUNDARIES

Why is it so hard to see 'women' and female bodies in practices that focus so much on female body parts? How can we understand that manipulating ovaries and wombs is compatible with talking about 'men' and 'children' as the individuals concerned and not 'women'? How is it possible that the notion of women as individuals is deconstructed in these practices, whereas the individuality of men and children appears to remain in place?

The answer to these questions is already partly contained in the analyses presented in the previous chapters. I will suggest here that it is precisely in "deconstructing" the notion of women as (embodied) individuals that these technologies can be designed as treatments for others, and, ultimately, that the individuality of the men and fetuses involved can be construed as unproblematic and stable in these practices. It is neither coincidence nor a natural necessity that women and women's bodies have disappeared from view in these new medical-technological discourses, but rather a built-in characteristic of these technologies, one that make them 'work' in the first place. As I will argue, it is a requirement or even an accomplishment of these technologies, rather than a necessity following from the nature of bodies or reproduction.

To develop my argument, a detour is needed first. Connected to the notion of an individuated or individual body is the concept of body boundaries. To be visible as an individual body, some sense of a boundary of that body has to be there. To see something as an individual entity, an Einzelkörper, a demarcation is required, marking what does and does not belong to that body. Body boundaries perform this function, defining the inside and the outside, self and non-self. Therefore, part of the answer to the question why women and female bodies go unrepresented in these technological discourses can be obtained by exploring the issue of female body boundaries. In the context under discussion, it is no longer clear what constitutes the female body proper, and thus what constitutes the body a woman may call 'her own'. However one may value the new possibilities for intervention, with the simultaneous creation of 'patients' like fetuses and couples, female body boundaries become rather fuzzy. The lack of representation of women as individuals may be connected to unclarities in the demarcation of their bodies from 'others'. Therefore, this chapter will set out to explore the role of body boundaries in these practices.

Seen in this light, the modus operandi in IVF and fetal surgery can be described as oriented in large part precisely toward the goal of overcoming body boundaries. Much of the work involved in both practices serves the purpose of rendering opaque bodies transparent. The goal of obtaining access underlies many of the interventions, and possibilities for intervention are created by externalizing processes and phenomena internal to the female body, overcoming the distinction between the inside and the outside of the body. Opening what is closed to intervention, disclosing what is hidden from inspection, getting out what resists easy manipulation - these are the recurring themes.

The prominence of ultrasound technology, both as 'diagnostic' apparatus and guiding instrument in invasive procedures is highly significant here, indicating the centrality of the need to overcome body boundaries and create transparency. It is used in monitoring follicle development, extraction of ova, and confirmation of pregnancies after embryo-transfers in IVF, and in fetal surgery it is a near constant presence providing visual access to the fetus, before, during, and after most procedures. Ultrasound and its ubiquitous use in modern medicine has been analyzed by many authors as a primary example - within a range of contemporary visualization technologies - of the way modern medicine transforms opacity of bodies into transparency.[257] In less general terms, it has been described how its first and foremost use has been in obstetrics, a fact from which its cultural significance is derived as a potent, political factor in changing general perceptions of pregnancy. Generation of a new and compelling iconography of the fetus with a broad cultural, political, and psychological impact has been attributed to visualization techniques.[258]

The description of modern (reproductive) medicine as aiming at externalization of inner processes and tending to unveiling what is hidden from view can be found in the discourse of reproductive medicine itself. Such metaphors are even conspicuously present in the following quote from M.R. Harrison, one of fetal surgery's leading figures:

> The fetus could not be taken seriously as long as he remained a
> medical recluse in an opaque womb; and it was not until the last half
> of this century that the prying eye of the ultrasonogram rendered the
> once opaque womb transparent, stripping the veil of mystery from the
> dark inner sanctum, and letting the light of scientific observation fall
> on the shy and secretive fetus. The sonographic voyeur, spying on the

> unwary fetus, finds him or her a surprisingly active little creature, and not at all the passive parasite we had imagined.[259]

This quote is taken from an article in which the author is explicitly reflecting on the developments in his field. The exuberance of the metaphoric language may therefore be attributed to a deliberate effort on the author's part at 'fancy writing'.[260] But when we turn our attention toward the more mundane types of scientific writing it becomes clear how the issues of visibility and access are explicitly put forward as underlying countless, very concrete choices and interventions. The following examples show that efforts toward permeating bodily boundaries and optimizing transparency determine choices as concrete as ways to suture surgical wounds and underlie curious measures like insufflating wombs.

> The maternal abdomen is then closed. It is important to use a subcuticular maternal skin closure covered with a transparent dressing so that monitoring devices can be placed on the maternal abdomen postoperatively.[261]

> Endoscopic fetal surgery uses a telescopic lens and operating instruments that are passed through small "ports" in the uterus. A bubble of CO_2 is used to displace amniotic fluid and provides excellent visualization in a magnified field.[262]

> For these reasons gas insufflation was used, as initial trials demonstrated excessive light scatter and distorted optics when visualizing through amniotic fluid. The air pocket creates a space in which surgical manipulation can easily be performed by displacing the uterine wall away from the fetus and allows for the effective use of cautery.[263]

A transparent wound dressing is applied so that skin remains permeable for ultrasound waves; holes ("ports") are made in uterine walls for telescopes and other instruments to pass; amniotic fluid is displaced by gas insufflation because of its "distorting" optic qualities.

In addition to the externalization of internal phenomena through graphic visual representation and the entering of bodies with means to visualize insides, the transparency of the female body is accomplished through other kinds of medical actions as well. Activities and interventions subsumed under categories like monitoring, surveillance, and data gathering have, in a more Foucauldian sense, similar effects. In the unrelenting search for knowledge about the body and its changing conditions, these activities make the body yield information about many aspects of its functioning. In IVF, there is constant monitoring of follicle development; hormone levels are measured to detect imminent ovulation and time ovum aspiration; the state of the endometrium may be measured before embryo transfer; presence of elevated HCG levels and gestational sacs are tested after embryo transfer in order to establish pregnancy. In fetal medicine an even greater variety of data is required, each serving to further "strip the veil of mystery from the dark inner sanctum, and let the light of scientific observation fall on the shy and secretive fetus" and "render the once opaque womb transparent." Partly depending on what anomaly is suspected, bodies are probed for information on heart rates, movements, temperatures, growth, in- and output of fluids, chromosomal or genetic make-up; they are made to provide samples of blood, amniotic fluid, fetal or placental tissue, etc. This is a more metaphoric form of creating transparency: gathering knowledge and information is another label for stripping bodies of their privacy and secrets; maximization of surveillance and monitoring requires a panopticon.[264]

Finally, both technologies encompass externalizations of inner processes in the most literal, dramatic,[265] and material sense of the word. In IVF, fertilization and early embryonic development are displaced from the female body and made into laboratory events; in fetal surgery in its most radical form, that is, open surgery involving hysterotomy, the fetus is (sometimes partly) "exteriorized", as it is literally called by practitioners.

Rendering body boundaries permeable in multiple ways, dissolving the distinction between inside and outside the body thus forms an constitutive element in the way these technologies work. However, this is not a characteristic particular to current developments in reproductive technologies. Though it may be that reproductive technology in its contemporary form is one of the most extreme instances, it should perhaps rather be considered a characteristic of modern medicine generally. When looking into discourses on heart surgery, for example, we will find the same scattering of bodies into bits and pieces, body boundaries will similarly be transgressed, and many of the efforts involved will be oriented toward gaining visibility of and access to what normally is hidden from view, exposing the inner body in comparable ways. Therefore, it is not the fact that women are reduced to "bits and pieces", to organs and cycles, that accounts for the disappearance of female individuality from this discourse. It is not "reductionism" per se that is inadvertently but unavoidably bringing this along, because the same kind of reductionism can be found in discourses and practices on appendicitis, heart disease, or broken legs.

The difference is, however, that in those cases there is no question about who this reductionalistically 'fragmented body' belongs to, or how many and which individuals are concerned when reference is made to 'the patient'. There will be one individual patient with one individuated body, and no 'others' will have been written into this patient body. There may be a body that was temporarily stripped of its bounded wholeness, but this all served the unequivocal purpose of making that individual body whole again.

The dissolution of body boundaries encountered here facilitates something quite differently: it replaces discontinuity between discrete, individuated bodies by continuity. In dealing with human reproduction, of course, there is not one, but there are, in a sense, always three bodies involved. As a process, reproduction starts with two bodies, the female and male ones, and results in a third body, the child's. By bracketing the body boundaries of one of these in particular, a continuum is created on which transitions (both in time and space) from the one to the other become very unclear. With 'the fetus' and 'the couple', technology has created twilight zones in which the individuality of bodies is suspended. In the cases of male infertility and congenital disease, this is done to the explicit purpose of overcoming medical problems in men and children, respectively. Thus, what the creation of fetuses and couples does, what the dissolution of female body boundaries makes possible, is, in effect, an extension and stabilization of the bodies, their boundaries, and the individuality of men and children, and of the range of medical care for them. The creation of a continuum between bodies by clearing away female body boundaries makes it possible to displace male body functions, pathology, and agency far beyond what common sense would consider male body boundaries. Similarly, by rendering a pregnant body transparent in all the senses of the word imaginable, it has become possible to extend a child's individual existence far back into what used to be unequivocally a woman's body. Unlike the case of heart surgery, therefore, the diffraction of bodies in bits and pieces, or the dissolution of these bodies' individuality encountered in these cases, serves the exact purpose of enabling the notion that these practices are about other bodies; other bodies whose individuality is produced as unproblematic in this very process.

3. SHIFTING BODY BOUNDARIES AND EXPANDING TECHNOLOGIES.

With 'the fetus' and 'the couple', ambiguous 'patients' have been created, that render the issue of what exactly belongs to whose body is, at least for the time being, irrelevant. This in itself is not necessarily of great concern; it becomes more problematic when we see how this temporary suspension of individuality subsequently leads to redefinitions of what belongs to each of the three bodies concerned that systematically move in a particular direction. Of the three bodies involved in reproduction, two are systematically gaining a broader definition: the scope of what is seen to belong to their individual functioning is extended, and so is the medical care for these bodies. The third, the one "in the middle", is, in one and the same movement consistently becoming less. Not enough to count as one whole individual, and be represented as such, it seems.

In contrast with the other two bodies, the female body is demarcated from 'others' beneath skin level. In both cases a difference between female and other bodies is discursively maintained, or, rather, redesigned. 'Maternal factors' are distinguished from 'fetal factors' within the "fetal-maternal-unit", female ones differentiated from male ones within "gamete interaction" and fertilization. Thus the boundary between female bodies and others' bodies seems to lie now at the level of physiological processes and interacting cells beneath the skin.

If we do not assume that body boundaries are naturally given but contingent, it becomes possible to see how such boundaries are construed, dissolved, and redrawn in relationship to developments in knowledge and technological possibilities, in mutual dependency with shifting definitions and relocalizations of medical problems. Once we take this view, it becomes also clear that nothing in principle will prevent such boundaries to shift once more. The way the immediate problems limiting success of fetal surgery and IVF for male infertility are identified suggests that further shifts are imminent. With the development of both technologies, the problems keep shifting, and the 'targets' for improvement or refinement of the techniques change accordingly. They often do so, however, in a particular direction. This results in an incremental process in which (the medical care for) the male body and the child's body are gradually extending, by incorporating ever more parts and aspects of the female body. This process did not stop at the skin, and, as we will see, neither will it stop at the next identifiable boundary, the one between "maternal" or "female factors" and others.

To describe this process, I will borrow a concept devised for quite a different technological context: the dynamics of technology development as described by Hughes (1987), in relationship to elecricity distribution systems. Hughes finds the direction of such a development often determined by 'reverse salients', a military metaphor that refers litterally to a lagging position on a front line. In the context of technological development it means "components in an expanding system in need of attention, such as a drag, limits to potential, emergent fricton, and systemic efficiency."[266] From the point of view of the technological system, factors not under control, uncertainties limiting success, constitute "environment". "Over time, technological systems manage increasingly to incorporate environment into the system, thus eliminating sources of uncertainty."[267] In order to improve technologies, elements of the "environment" are brought under control, causing the expanding technology to turn former "external factors" into system components.

Without wanting to take up a systems-approach here, and however the two contexts may differ otherwise, some of its concepts prove surprisingly suited for describing the role of the female body in reproductive technologies. As many analysts of contemporary prenatal medicine and embryology have observed, from the point of view of the search for control over health of the fetal or child's body, the female body came to be conceptualized as "environment"[268]. In line with Hughes's description of the dynamics of expanding technologies, many of the problems of fetal medicine subsequently came to be located in this by definition 'uncontrollable' body-as-environment, the "maternal" factors "external" to the fetal body. Hence, the development of fetal surgery follows the path of seeking control over these factors, and transforming them from external into internal components of fetal/child health.[269]

A similar story can be told for IVF and male infertility. As we have seen, via the notion of "gamete interaction", male infertility became step by step transformed into a problem caused by factors external to the male body, in relation to which the female body and "female factors" came to be perceived as uncontrollable and hindering environment. Agency in fertilization is still squarely attributed to male gametes. Though theoretically, a failed fertilization may be attributed to low sperm quality, the pragmatics of the technology turns this around in focusing on "female factors" in order to overcome the problem. The accompanying discourse rationalizes this reversal in actually speaking of hindrance and withholding of opportunities by the female parts. Following Hughes's logic of the military metaphor of reverse salients as determinants of the direction of technological development, this conceptualization is consistent with subsequent efforts of "improving" IVF as male infertility-treatment by directing efforts at other female factors than standard IVF implies. Examples include innovations such as micromanipulation of oocytes, tubal embryo transfer (TET), supplemental hormonal treatment after embryo transfer; these are all examples of gradual incorporation of more and more "female factors" into the treatment of male pathology. Thus the technological efforts to solve the problems of male infertility proceeded by including female body factors in the domain under control, turning the latter from external, or environmental factors into controllable, integrated components of the male reproductive body.

Let me illustrate this mechanism in some more detail by tracing one particular "reverse salient" in fetal surgery. According to many of its practitioners and advocates, the most pressing problem for fetal surgery lies in the "uncontrollability" of the uterine contractions resulting from surgical manipulation of the sensitive pregnant womb. These contractions often start as soon as the surgery on the womb begins, and, if not controlled, usually result in premature labor and/or perinatal death.

> Appropriate surveillance and treatment for preterm labor remains the "Achilles heel" of fetal surgery. Ineffective treatment of preterm labor led to delivery at 26 weeks gestation.[270]

> Indeed intraoperative contractions caused significant problems in at least half of these cases, and clearly ruined an otherwise successful repair in two.[271]

> Finally, even when all these technical maneuvers were accomplished (case 11), success was spoiled by uncontrolled uterine contractions which began as soon as we made the hysterotomy, caused repeated bradycardias throughout the procedure, and did not respond to deepening isofluorane anesthesia.[272]

This way of framing the problem of the often disastrous results of surgical intervention is clear on where the "blame" is to be put: the concept of intervention as well as the technical activity themselves are sound, but the uncontrolled contractions "spoil" and "ruin" the "otherwise successful" interventions. Again, the issue limiting success of the technology is framed as a lack of control over external, that is (normal, or at best, iatrogenic) female physiological, factors. The reverse salient posed by the uterine contractions, as a 'maternal factor', is literally conceived of as a "limit" and a "barrier" to be overcome:

> Intraoperative technical problems have been overcome; the factors limiting successful outcome are postoperative physiologic management of the maternal-fetal unit and effective tocolysis to control preterm labor.[273]

> ...the one great remaining barrier to fetal intervention: the sensitivity of the human uterus and the threat of inducing preterm labor and abortion.[274]

And, reminiscent of the "final frontier" in American mythology,[275] this boundary symbolizes challenge to further exploration rather than an actual limit.[276] This "barrier" to intervention on behalf of the fetus is to yield by extension of the domain under control with an other part of maternal physiology. The only way thus far to deal with the (iatrogenic) problem of uterine contractions, however, is the administration of so-called "tocolytic therapy", a type of medication that is notoriously inadequate in these cases.

If the pregnancy is to be saved beyond the surgery, the women usually have to remain on this medication for the rest of their pregnancy, which may be up to several weeks or even months, but in many cases become so ill from it that continuation becomes dangerous for them. It is even labeled "toxic" by some doctors for the common incidence of quite serious side effects:

> The standard obstetrical regimen for tocolysis was inadequate and even toxic. ...Case [X] had a very difficult postoperative course due to poorly controlled labor and the toxicity of magnesium sulfate and terbutaline.[277]

However:

> This *final barrier* may now yield to the recent demonstration that the nitric oxide-cyclic GMP pathway is involved in uterine relaxation, and can be manipulated therapeutically.[278]

But as long as this particular barrier refuses to yield, alternative routes are devised based on a slight, though familiar reformulation of the problem: the issue is once more framed in terms of the problematic opacity of the female body. Inadequate monitoring, lack of post-operative accessibility of the fetal patient, and lack of information concerning its intra- and postoperative well-being push developments in the direction of enhancing transparency once more. This time with an emphasis on making it last over time and space, giving it an even stronger panopticon-like quality.

> Even when fetal deterioration was accurately detected (case 13), attempted fetal resuscitation was hindered by our inability to access the fetal circulation. Recently we have developed endoscopic catheterization of placental vessels for chronic vascular access.[279]

> Bradycardia can progress quickly to fetal demise. The radiotelemeter ECG solved the postopertative as well as intraoperative monitoring problem. ... Monitoring of uterine activity by external tocodynamometer was inadequate to allow adjustment of tocolytics. This problem has been solved experimentally by using the radiotelemeter to monitor uterine electromyeogram and intrauterine pressure.[280]

> The implantable radiotelemeter may have a major impact on postoperative fetal monitoring. ... In addition, this system may allow us to monitor the fetus while the mother is at home via custom interface to a standard ECG audio modem transmitter and receiver system.[281]

> In the future we would like to develop a simultaneous real time display of fetal temperature, ECG, and fetal heart rate variability, pH and oxygenation status, as well as uterine electrical activity and

> intrauterine pressure. The postoperative fetus can no longer be thought
> of as locked away and inaccessible with inadequate monitoring and
> few therapeutic options, but rather as being in a uterine "fetal intensive
> care unit".[282]

It must be born in mind here, that the problem of induced labor from surgery and morbidity from tocolytic therapy is primarily conceptualized as a problem for the fetus: it is "fetal outcome" that is being threatened by these problems. It may be that such problems are experienced in different terms in clinical encounters; as a medical-scientific problem it is conceived in terms of "fetal well-being". Accordingly, solutions to these problems are primarily conceived of as improvements in fetal treatment. This means, again, that all further changes in the approach and manipulation of the 'maternal body' become included in "management of the fetus". Thus a gradual extension of the fetal/child's body is accomplished by technological inclusion of ever more elements of "maternal physiology" in its care.[283] By implication, the boundaries demarcating what unequivocally constitutes the female body, as opposed to someone else's body, have shifted as well.

4. THE POLITICS OF ONTOLOGY / LIVING BODIES AS PROSTHESES

If the problem-definitions in fetal medicine and male infertility treatment turn the female body into environment, the technological solutions to those problems seem to transform it into a kind of prosthetic device for the fetal/child and the male body. A prosthesis is an artificial device that replaces a missing or dysfunctional part of the body; it compensates for a lost bodily function or capacity. Furthermore, it is purely instrumental, infinitely adaptable, malleable and manipulable so as to reach an exact fit with the incomplete body whose dysfunction it is to compensate. It is attached to that body, is part of it, defined in relation to it, yet distinguishable from it.

The mentioning of the "uterine intensive care unit" indicates that it is not too far-fetched to suggest that the female body is at times imagined as a set of multifunctional and usable equipment - to be put to good use for another body.[284] Such comparisons appear regularly, and in a variety of forms. Next to the (transparent) womb as intensive care unit, we find, for example, the maternal blood system and placenta as "ECMO", as a "by-pass", the womb as "isolating bubble" (as used for patients with dysfunctional immune systems), "the perfect heart-lung-machine", "the ideal operation room" or "recovery room"[285].

> Long term support or replacement of lung function after birth will
> require either an artificial placenta or neonatal lung transplantation.